RAT TRAP

RAT TRAP

THE CAPTURE OF MEDICINE BY ANIMAL RESEARCH AND HOW TO BREAK FREE

PANDORA POUND, PHD

Copyright © 2023 Pandora Pound
Cover image © Shutterstock
Author photograph © Kristy Field

The moral right of the author has been asserted.

Apart from any fair dealing for the purposes of research or private study, or criticism or review, as permitted under the Copyright, Designs and Patents Act 1988, this publication may only be reproduced, stored or transmitted, in any form or by any means, with the prior permission in writing of the publishers, or in the case of reprographic reproduction in accordance with the terms of licences issued by the Copyright Licensing Agency. Enquiries concerning reproduction outside those terms should be sent to the publishers.

Troubador Publishing Ltd
Unit E2 Airfield Business Park
Harrison Road, Market Harborough
Leicestershire LE16 7UL
Tel: 0116 279 2299
Email: books@troubador.co.uk
Web: www.troubador.co.uk/matador
Twitter: @matadorbooks

ISBN 978 1 80514 052 8

British Library Cataloguing in Publication Data.
A catalogue record for this book is available from the British Library.

Printed and bound by CPI Group (UK) Ltd, Croydon, CR0 4YY
Typeset in 11pt Adobe Garamond Pro by Troubador Publishing Ltd, Leicester, UK

Matador is an imprint of Troubador Publishing Ltd

To my mother and father
Joanna and Pelham Pound

'It's a rat trap ... and we've been caught'
Bob Geldof, 1978

CONTENTS

Preface ix

Part One: Trapped 1
1 Capture 3
2 Stuck 20

Part Two: In captivity 35
3 Bias at the bench 37
4 Elephant in the lab 56
5 Vital and indispensable? 68
6 Desperate patients 83
7 The real guinea pigs 96

Part Three: Breaking free 111
8 The potential of a human cell 113
9 Unleashing the power of computers 129
10 Something new, something old 142

Part Four: The struggle to move forward 161
11 Regulatory dysfunction 163

12 Locked in	181
13 Death throes and birth pangs	194
Acknowledgements	216
End Notes	218
Index	274

PREFACE

During the many years I worked in medical schools, I was always aware that somewhere in the depths of these vast institutions was a laboratory housing animals for use in medical experiments. My own research never brought me into contact with the scientists who inhabited these laboratories, and the only person I knew who worked in one, a secretary, seemed uncomfortable when I asked her about it.

'It's necessary for medical progress,' she said warily.

In the late 1990s, I was part of a team conducting research into stroke, working with doctors, statisticians, epidemiologists and other social scientists at St Thomas' Hospital Medical School in London. The hospital and medical school occupied a site in the shape of an elongated triangle and my office was at its bottommost tip, where Lambeth Palace Road meets the River Thames. Each morning I would cross Westminster Bridge, walk through the ground floor of the busy hospital and then through Block Nine, which housed the medical school. It was a long walk to my office and the further I went, the fewer people I met, until eventually I found myself in an empty corridor. Stretching into the distance, this corridor had several red doors along the right-hand side which were always locked and, unlike other doors within the medical school, bore neither nameplates nor room numbers. At last I emerged through a set of fire doors into a little yard surrounded by a high brick wall. The building that housed my office was across this yard.

One day I heard a commotion in the yard and, peering out of the window, saw a sheep being unloaded from a van into a narrow crate. I was confused; what on earth was a sheep doing here? A woman looked up, saw me watching and quickly turned away. Suddenly it dawned on me where that sheep was heading and what lay behind the red doors.

Shortly afterwards I visited a colleague in the main department, a gloomy collection of offices that the sunlight never seemed to reach. Almost subterranean, the department was crossed by a service tunnel that ran beneath the medical school. As I entered, a man wearing a lab coat emerged from the tunnel, pushing a trolley bearing a partially covered cage. He walked past me and disappeared once again into the tunnel, heading in the direction of the red doors.

***

Although animal research is still secretive, at least some conversations are now possible. In those days there was simply no dialogue between the scientists conducting animal experiments and people who might have questions about the practice; the latter were simply dismissed as anti-vivisectionists. I have never conducted animal experiments – my background is in the sociology of health and medicine – so as an outsider, it was hard for me to raise questions about the utility of animal research or its underpinning science. I had seen a request for lay members to join an ethical review board overseeing animal studies within the medical school, so I contacted the chair in the hope that this might enable me to find out more. Our conversation was brief. On hearing that I had some questions about the science, he told me in no uncertain terms that I was not the sort of person he wanted on the board.

In 1998, the medical school moved to another location and Block Nine was abandoned. In 2015, the derelict site was investigated by urban explorers who photographed what they found. In the long corridor the red doors now swung open. Plastic pots containing samples of rat organs littered the surfaces of offices, scattered alongside empty boxes and vials of Hypnorm, an anaesthetic commonly used in experimental surgery. Peeling paint hung in swathes from the ceilings of empty laboratories, and the semicircular bench in the lecture theatre was now bright green with moss. The service tunnel stretched into blackness and in dim, windowless spaces, the gates of barren steel cages and pens stood open, some still containing the animals' feeding bowls. According to an ex-student who commented on the photographs,[1] some of the larger cages had housed monkeys and sheep. I was left wondering what, if anything, the animals who lived in those dark cages had contributed to human medicine.

This question intrigued me over subsequent years. In 2004, and working in a different medical school, I attempted to address this issue in a *British Medical Journal* article. The paper, written with four professors of epidemiology, was plainly titled

'Where is the evidence that animal research benefits humans?' and simply argued that claims about the value of animal experiments needed to be supported with data.[2]

The reaction was horror and dismay. The paper was described by one critic as 'spectacularly ill-judged' and 'scientifically invalid', while another stated that it 'should never have been published in a peer-reviewed journal'.[3] Mark Henderson, now Director of Corporate Affairs at the Wellcome Trust, attempted to publicly discredit us in *The Times*, calling us 'the anti-vivisection lobby, or at least its law-abiding element'.[4] In *The Telegraph*, the lobbying group Coalition for Medical Progress (now known as Understanding Animal Research) protested that animal experiments had led to advances, citing polio vaccines, kidney dialysis, stomach ulcers and cystic fibrosis.[5] On the day our paper was published, the UK's Royal Society published a 'guide', the opening lines of which claimed, 'Humans have benefited immensely from scientific research involving animals, with virtually every medical achievement in the past century reliant on the use of animals in some way'.[6] Professor Colin Blakemore, then Chief Executive of the Medical Research Council, publicly backed the Royal Society's position, asserting, 'Animal research has contributed to virtually every area of medicine'.[7] Clearly, the scientific establishment was rattled; in 2004, animal research was still a 'sacred cow' and it was considered outrageous to question its benefits.[8]

Yet this is surely a legitimate line of questioning given that we still have very little to offer those suffering from common diseases such as stroke, Alzheimer's disease, dementia and cancer, despite decades of animal research. A shocking ninety per cent of all experimental drugs fail in human trials despite having first passed tests in animals, and in many fields the statistics are even worse.[9] Moreover, despite animals being used to test the safety of medicines, alarming numbers of people suffer from serious, sometimes fatal, adverse drug reactions both during clinical trials and after drugs are approved.[10] The arthritis drug Vioxx, for example, is estimated to have caused between 88,000 and 140,000 excess cases of serious coronary heart disease in the US alone, many of which were fatal, before it was removed from the market.[11] So in addition to asking whether animal research leads to beneficial new drugs, we need to question whether it protects us from harmful ones.

In the intervening years, animal experiments have come under unprecedented scrutiny, and it is now abundantly clear that the vast majority are not conducted according to accepted scientific standards.[12,13,14,15] It might be thought that in such a morally contested field, scientists would take particular care to conduct their research properly in order to produce useful findings and make each animal's life count. Unfortunately, for the most part, this is not the case; review after review has revealed the quality of animal studies to be poor, meaning that the findings cannot be trusted. This is an open secret within scientific circles.

In 2017, I attended a meeting about animal research at the University of Bristol where I was working at the time. The event was unusual because it was open to all staff members and was probably the first such meeting about animal research to be held there. Having recently signed the Concordat on Openness in Animal Research,[16] the university was perhaps attempting to engage with staff on the topic. The Concordat had been launched by Understanding Animal Research, a UK group representing the interests of scientists using laboratory animals, ostensibly to enable members of the public to find out more about their work.

'How do you think members of the public might react if they discover how poorly animal research is conducted?,' I asked when it was time for questions.

The representative replied to the effect that these were technical issues that the public wouldn't understand or be interested in. Yet I believe people *are* interested in knowing just how poorly conducted and unscientific animal research often is and just how careless scientists can be with animals' lives. Poor quality research generates untrustworthy and misleading findings, meaning that the research is completely wasted. So, should efforts focus on improving the quality of animal research?

This book argues that such an approach is unlikely to reap rewards because the quality issues are eclipsed by a much larger issue: the differences between animals and humans. Each species is unique, and even the subtlest differences can have significant consequences in terms of responses to pathogens and drugs, as we shall see. If the whole enterprise is beset by this fundamental flaw, there seems little point in attempting to improve the quality of animal experiments. It would be a bit like touching up the paintwork on a car that has no engine. So, does this car have an engine? I consider whether animal research drives medical innovation, as is frequently claimed, and investigate the extent to which it generates treatments for humans and ensures the safety of new medicines.

***

PREFACE

Imagine you are an alien visiting our world with an assignment to investigate how we conduct research into human diseases and treatments. Strangely, you discover that a huge proportion of this research focuses on animals rather than humans. Suddenly, you become ill with a mysterious disease. You hope that scientists will examine you carefully, listen to what you say about your symptoms, scan your organs and analyse samples of your cells and fluids using the latest technologies. You hope that they will try familiar, approved drugs with known mechanisms of action and proven safety profiles, or test new drugs using advanced technologies that incorporate living human cells and simulate human organs. But no, astonishingly, they try to manufacture the disease in mice, rats and monkeys, test their experimental treatments on these animals and then try those treatments on you!

There is no longer any excuse for such an approach. Today, technology allows us to examine organisms in great detail. We now know that differences at the genetic, molecular, cellular and tissue levels can have profound consequences for the ways different species develop diseases and respond to pathogens and drugs. Thanks to recent scientific and technological advances, we now also have a range of tools at our disposal for understanding and generating knowledge about the human body and how to treat it. These new research approaches and cutting-edge technologies are grounded in *human* biology so have direct relevance to us. Inspiringly, they have the potential to both replace animals *and* accelerate medical progress. Most stakeholders in this debate agree that replacing animals with twenty-first-century scientific methods is the ultimate goal. An important theme of this book, then, is to illustrate that this is now scientifically possible. We will explore the exciting new realms of in silico modelling, systems biology, organoids and organs-on-chips. In the process, it will become clear that our continued use of animals has more to do with custom than with any scientific imperative.

Ten years after the publication of our paper that so incensed the animal research establishment, we published a follow-up paper, also in the *British Medical Journal*. Here we reviewed the evidence that had accumulated in the intervening years, concluding that animal studies suffer from serious limitations and produce few clinical benefits.[17] This time, our paper was featured on the front cover. We were told that it was 'much appreciated by those of us attempting to improve biological sciences' and thanked for 'doing the medical establishment a service by highlighting that the approach in general may be broken'.[18] The sunny reception showed that it was no longer taboo to question the validity or legitimacy of animal research, illustrating a shift in the scientific landscape.

Interestingly, the ethical landscape is shifting too. More people are now showing an interest in the welfare of laboratory animals and in animal-free methods of research.[19] Recognition of animals' capacity to experience feelings is also growing, both generally and within the law;[20] in 2022, the Animal Welfare (Sentience) Act became UK law, meaning that the sentience and welfare of vertebrate animals must be considered in any new legislation. Meanwhile, people all over the world are withdrawing their support for animal research. In the US, fifty-two per cent of the public now oppose the use of animals in scientific research,[21] while over two-thirds (sixty-eight per cent) of British adults want an end to animal experiments for medical research[22] and sixty-six per cent of EU citizens want all animal testing to end immediately.[23]

Clearly, there is a thirst for change. The Netherlands has a strong programme to accelerate scientific innovation without using laboratory animals,[24] as does the US Environmental Protection Agency. And in September 2021, the European Parliament voted by a stunning majority of 667:4 to develop a coordinated plan to replace animal experiments with innovative, non-animal methodologies.[25] However, there are also signs of a concerted and deliberate pushback against these developments. After a century and a half of animal research, the practice is now baked into our institutions, our economy and our culture, and those whose careers and livelihoods depend upon it are fighting hard to preserve the status quo.

Many books have been written on animal research, but most have been about ethics and the impact on animals. At one end of the spectrum, laboratory animals experience boredom, frustration and the stunting of their instinctive behaviours. At the other, they suffer unimaginable and prolonged agony. I recognise this, but *Rat Trap* does not focus on animal suffering. Rather, it concentrates on the scientific method as it relates to animal research, the impact of the practice on humans and the new methodologies available to replace it. In what follows, we will draw upon the latest evidence to scrutinise the science of animal research, explore the breathtaking potential of new technologies and examine some of the barriers to change. In support of the scientific arguments I marshal the best available evidence, which generally means peer-reviewed reports of research and syntheses of bodies of research where these are available. I include my own experiences, as well as those of experts I have interviewed, from fields as diverse as medicine, evolutionary biology, biotechnology and regulation.

I focus on the use of animals in the development and testing of pharmaceutical drugs, not because I consider drugs to be the most important way of ensuring human health, but because this is the arena in which the use of laboratory animals is most firmly embedded. I do not consider surgical or other medical interventions that might be developed using animals, nor the use of animals in research outside medicine (e.g. for industrial chemicals, cosmetics or weapons).

A number of factors are contributing to the changes that are beginning to take place in the way we develop and test drugs, but first and foremost it is the failure of animal research. As Thomas Kuhn, the philosopher of science, wrote, 'scientific revolutions are inaugurated by a growing sense … that an existing paradigm has ceased to function adequately in the exploration of an aspect of nature to which that paradigm itself had previously led the way'.[26] The science of drug discovery is undergoing dramatic change, and a battle for its future is underway, yet few are aware of the dramatic developments or of what is at stake. *Rat Trap* attempts to convey a sense of this approaching revolution.

PART ONE

TRAPPED

1

CAPTURE

On 22 August 1870, Marie Françoise Bernard divorced her husband, the physiologist Claude Bernard. It cannot have been easy at that time for a French woman to obtain a divorce, particularly as a Roman Catholic, so what impelled her to take this step?

The Bernards' had been an arranged marriage, with Marie Françoise's dowry helping to fund the work that made her husband famous: experiments on living animals. Marie Françoise was deeply unhappy about her husband's activities, some of which apparently took place in the cellar of the family home.[1] After twenty-five years of marriage, she had had enough. She left with her two daughters, and all three became passionate activists in the field of animal protection.

Claude Bernard, meanwhile, went on to become a pivotal figure in the history of animal experimentation, his influence extending well into the present day. Understanding the reasons why he and his colleagues conducted animal experiments, as well as how they responded to their opponents, is key to appreciating how the practice came to exert such a grip upon the biomedical sciences. For it has certainly gained a remarkably strong position. Today, animal research is conducted within medical schools, university laboratories, pharmaceutical companies, commercial facilities and military research establishments all over the world, with an annual estimate of up to 192 million animals used in science worldwide.[2] With such widespread use and the practice so firmly embedded within our institutions, we might be forgiven for assuming that it is indispensable for understanding human biology and vital for developing and testing treatments for humans. Unfortunately, as we shall see, the evidence does not support this view. So what is going on? How did the practice manage to establish itself so successfully?

The rise of animal experimentation

Prior to the nineteenth century, animals were used in experiments only intermittently. Hippocrates, the fifth-century BCE Greek physician, preferred to rely solely on the careful observation of humans, but a century after him, the Greek philosopher Aristotle was one of several prominent physicians to conduct dissections and experiments on animals and to influence the physician Galen to do similarly in the second century CE.[3] Human autopsies were forbidden in Galen's time, so his conclusions about human anatomy were formed on the basis of animal – particularly ape – dissections, contributing to a belief that the anatomy of humans and animals was essentially similar.[4]

Although interest in medical knowledge and vivisection[*] declined as Christianity, with its focus on spiritual explanations of disease and healing, took hold throughout the Middle Ages,[3] the Renaissance brought a renewed curiosity in the functioning of living organisms. The sixteenth-century Flemish anatomist Andreas Vesalius conducted dissections and vivisections,[5] comparing human anatomy with that of other animals, while in the following century, the English physiologist William Harvey manipulated the hearts of living animals to investigate blood circulation and heart function. Animals began to be used to investigate biology and pathology more frequently, and the theory of the seventeenth-century French philosopher René Descartes, that animals were unfeeling 'automatons', may have persuaded hesitant scientists to practice vivisection at a time when anaesthesia was unavailable.[3] On the whole, however, concerns about pain seem unlikely to have been a significant factor in scientists' attitudes towards the practice, since a considerable proportion of experiments continued to be conducted without anaesthesia even after its advent in 1846. Indeed, the 1873 *Handbook for the Physiological Laboratory*,[6] a manual for those conducting animal experiments and the first of its kind in the world, did not even contain a chapter – or indeed articulate a clear policy – on anaesthesia.[7] So we must look elsewhere to explain why vivisection became increasingly popular among scientists in the latter half of the nineteenth century.

---

[*] As the practice of experimenting on animals was commonly referred to as 'vivisection' throughout history, I will occasionally use this term here, switching to the more contemporary 'animal research' in subsequent chapters.

The emergence of the medical sciences

Historians agree that the rise of vivisection in the late 1800s was closely allied to the professionalisation of experimental physiology and its emergence as a distinct academic discipline.[8,9] In the early 1800s there was no established career path for experimental physiologists – those who use experiments to understand how living organisms work – so it was common to approach the subject by studying medicine or veterinary medicine. But career prospects were poor, and the paucity of academic posts and funding meant that many ended up migrating to other areas of science or medicine. Those physiologists who persisted frequently had to practise medicine to fund their studies. One such was the French physician François Magendie, born in 1783, who further supplemented his income by giving demonstrations of experiments on living animals to medical students, for which he became notorious.

To the advantage of physiologists, however, enthusiasm was building, particularly in France, about the experimental method as a basis for drawing scientific conclusions, and in 1821, the elite Académie des Sciences established the Prix Montyon as a new cash prize for contributions to research into experimental physiology.[9] Magendie attracted the attention of the Académie and, in the same year, was elected to join its distinguished membership. Ten years later, he was given the Chair of Medicine at the prestigious Collège de France and, once he had altered the title of the chair to *Médicine Expérimentale*, was able to focus all his attention on experimental physiology.

Eventually, Magendie became tutor to a young intern named Claude Bernard who would go on to become, in

1847, his deputy professor at the Collège. Between 1845 and 1853, Bernard won the Prix Montyon four times, supplying him with a generous income for conducting his experiments and, in 1854, was elected as a full member of the Académie. In the same year, he went on to secure the first chair in general physiology at the Sorbonne, and when Magendie died a year later, Bernard took over his chair at the Collège de France. He built on his mentor's approach, arguing that only properly controlled and rigorously conducted animal experiments could provide reliable information on physiology and pathology. His 1865 book, *Introduction à l'Etude de la Médecine Expérimentale*, set out the case for experimental physiology as an independent academic research discipline.[10] The key to achieving this status, he believed, was vivisection. Only this would raise physiology to the level of a 'true' science.[9]

Bernard and his colleagues were successful. The laboratory setting and the use of animal experiments succeeded in conferring a more scientific veneer upon experimental physiology, increasing its status, boosting career prospects and enabling it to compete with disciplines such as physics and chemistry.[8,9] Historians of science have not found it hard to conclude that the French physiologists' enthusiasm for vivisection was motivated less by a desire to understand disease, although Magendie and Bernard always used this as a justification for their work,[9] than by 'career ambitions and social aspirations'.[8] Indeed, the nineteenth-century novelist Maria Louise Ramé scoffed at the notion that physiologists were conducting animal experiments to advance medicine. 'Throughout Europe and America', she wrote, 'and in many parts of Asia and Africa, the pursuit of physiology is a profession like any other, a career, a means to an end – that end, like other men's, being money, celebrity and success'.[11]

Animal experimentation had a similarly strong influence on the status of medicine. During the early nineteenth century, medicine was practised at the patient's bedside, in their homes or in hospitals. Medical students learned physiology from surgeons,[9] and doctors developed hypotheses about the health, illness and treatment of their patients based on careful observation. Bernard held strong views on medicine, although he had never actually practised it. He argued that merely observing and predicting the course of natural phenomena was insufficient; one needed *power* over natural phenomena, and this, he proposed, was provided by the experimental sciences.[9] Medicine should be conducted in the laboratory, he declared, where experimental evidence could be collected and hypotheses tested, rather than at the patient's bedside or in hospitals. Animals would be the stand-ins for humans, who could not be experimented upon for moral reasons. Studying animals under controlled laboratory conditions and then extrapolating the results to humans would be a more scientific approach to medicine than directly observing humans or dissecting their bodies. The true sanctuary of medical science was the laboratory, he wrote in *Introduction à l'Etude de la Médecine Expérimentale*: 'only there can [the physician] seek explanations of life in the normal and pathological states by means of experimental analysis'.[10]

Bernard's views captured the zeitgeist. Several large medical research institutes focusing on laboratory-based medical research, including the Robert Koch Institute and the

Institut Pasteur, were established during the late nineteenth century, and animal experimentation assumed a central role within these new centres of learning. Medical practitioners, who were already attempting to improve the status of their profession with reforms, such as the requirement to register and meet elevated training standards,[12] benefited greatly from their association with the increasingly scientific character of medicine.[8] From the late nineteenth century onwards, animal experimentation and medicine became inextricably linked – as they still are – with the former contributing considerably to the rise of medical science as a respectable academic discipline.[13,14]

Rejection of the theory of evolution

Six years before Bernard published his book on experimental medicine, the English naturalist Charles Darwin published his groundbreaking work, *On the Origin of Species*,[15] the product of decades of careful research and painstakingly collected evidence. Darwin's theory shocked Victorian sensibilities because he proposed that all species descended over time from common ancestors. The religious establishment found his theory blasphemous since it meant that humans were no longer the pinnacle of a divinely created hierarchy in which species were fixed and unchanging. Darwin theorised that species gradually transform over time and that those possessing a variation enabling them to thrive or adapt, live to pass on their improved trait to the next generation. According to Darwin, then, each species is the unique, evolved product of natural selection, a theory that remains the central organising framework of biology.[16]

At its time of publication, however, Darwin's book received a mixed reception among scientists. Although many appreciated its message, not all accepted the new theory. Some vivisectionists used Darwin's findings on the common ancestry of animals and humans – the principle of homology – as justification for using the former to understand the latter, emphasising the similarities rather than appreciating the *uniqueness* of each species as Darwin's theory emphasised.[17] Bernard, however, rejected the theory as speculative and not properly scientific, believing it could not be tested experimentally. He was not unusual in rejecting evolutionary theory; it would take time to become widely accepted, even within biological circles. Yet the fact that Darwin's and Bernard's books were published within a few years of each other is startling. Their positions could not have been more different.

Bernard was of course aware of the differences between species, but his view was that all animals shared the same 'vital units', with species only differing in terms of how these units were arranged. It was this assumption that undergirded his view that findings in animals would be relevant to humans. Furthermore, Bernard regarded species differences as *quantitative* rather than qualitative, facilitating his belief that findings could be extrapolated from one species to another simply by making mathematical adjustments for differences in factors such as body weight, metabolic rate and surface area. Bernard's denial of the importance of the qualitative differences between species allowed him to pursue his notion that animals could be stand-ins for humans and would send the biomedical sciences down the path on which they continue to wander today.

Objections on grounds of poor science and futility

Interestingly, objections to vivisection on the grounds of species differences had been raised as early as the eleventh century, when the Persian physician Ibn Sina had strongly cautioned against extrapolating from animals to humans, emphasising, 'Experiments should be carried out on the human body'.[18] And in the seventeenth century, Jean Riolan Jr, Professor of Anatomy and Botany in Paris, stressed the significance of the anatomical differences between humans and animals, as did the influential philosopher-scientist Francis Bacon.

Riolan also argued that vivisection gave rise to completely unnatural conditions in animals, potentially causing their bodies to behave differently from humans.[8] Such objections, about the soundness of conclusions drawn from experiments which inflict extreme pain and trauma, had also been voiced in the third and fourth centuries BCE by the Empiric school of medicine in Alexandria. The school rejected the study of anatomy and physiology by vivisection and dissection, believing that pain and death distorted the normal appearance of internal organs and cast doubt on the results of animal experiments.[3]

As vivisection became more common in the nineteenth century, so did questions about whether it was justifiable in terms of the knowledge gained. While some scientists recognised that it contributed knowledge about the physiology of animals, others were excoriating in their critique of its relevance to humans, with one asking, 'Have we learned to treat any of the diseases of the nervous system more successfully, after all the horrible butcheries that have been committed within the course of the last five and thirty years, than did the wise and able men of last century, the Boerhaaves, the Cullens or the Heberdens? Was Sydenham, may we ask, educated in the school of blood and torture?'.[19]

There was also vehement *public* opposition to vivisection in France, Britain and elsewhere, on the grounds of cruelty and futility. Prior to 1870, relatively little vivisection had been conducted in Britain, but this was about to change. In 1873, the physiologist John Burdon-Sanderson's *Handbook for the Physiological Laboratory* was published,[6] symbolising the emergence of the new discipline in Britain and alerting anti-vivisection organisations to the extent of the practice. Considerable public disquiet on the matter resulted in the government establishing a Royal Commission on Vivisection in July 1875, which went on to recommend legislation to control the practice. The Cruelty to Animals Act of 1876, the first law in the world to regulate animal experimentation, allowed experiments to be conducted as long as their aim was 'the advancement by new discovery of physiological knowledge or of knowledge that will be useful for saving or prolonging life or alleviating suffering'.[20] Under the Act, a system of licensing experiments was created, with the Home Secretary overseeing the registration and inspection of laboratory premises and imposing penalties for offences.[21] Opponents described it as a 'vivisector's charter', partly because it permitted some experiments to be conducted without anaesthesia if this was deemed necessary for the experiment's success,[21] but also because it enabled the vast majority of experiments to proceed unhindered.[7] The Act, however, sent shock waves throughout the scientific establishments of Europe and beyond.

Pushback and the consolidation of vivisection

The first meeting of The Physiological Society, formed in the same year as the Act, was held in the parlour of John Burdon-Sanderson's house on Queen Anne Street in London. Nineteen men attended this meeting, with the Society initially established as a dining club.[17,22] The formation of the Society is commonly interpreted as a reaction to the Cruelty to Animals Act and as an attempt to protect and bolster the professional interests of physiologists. Specifically, it aimed to defend vivisection against attacks from its opponents, and one of its key plans was to launch a pro-vivisection campaign, in the form of a resolution on animal experiments, at the forthcoming International Medical Congress.[17]

Held in London in 1881, the Congress was 'arguably the largest and grandest medical congress ever held'.[14] Among the more than 3,000 participants were many famous scientists and physicians, including Rudolf Virchow, Louis Pasteur and Robert Koch. Several of the speakers at this huge conference made repeated assertions that vivisection was a necessary and justifiable means of advancing medicine,[14] and before the meeting closed, a resolution was passed unanimously, stating, 'that this Congress records its conviction that experiments on living animals have proved of the utmost service to medicine in the past, and are indispensable to its future progress; that, accordingly, while strongly deprecating the infliction of unnecessary pain, it is of opinion, alike in the interests of man and of animals, that it is not desirable to restrict competent persons in the performance of such experiments'.

Two days later, at its annual meeting, the British Medical Association issued its own resolution: 'that this Association desires to express its deep sense of the importance of vivisection to the advancement of medical science, and the belief that the further prohibition of it would be attended with serious injury to the community, by preventing investigations which are calculated to promote the better knowledge and treatment of disease in animals as well as man'.[14]

These public statements, clearly intended to ward off any further interference or restrictions to scientific freedom, indicate just how threatened scientists felt by the public opposition which culminated in the 1876 Act. Then, as now, they claimed that any infringement on their perceived right to conduct animal experiments would hinder medical progress. When, in the same decade, Anna Kingsford, one of the first English women to obtain a degree in medicine, asked why the *faculté de médicine* insisted upon vivisection, her tutor, Professor Leon Le Fort, replied, 'Speaking for myself and my brethren of the *faculté*, I do not mean to say that we claim for that method of investigation that it has been of any practical utility to medical science or that we expect it to do so. But it is necessary as a protest on behalf of the independence of science as against interference by clerics and moralists.'[23]

According to Le Fort, then, the ability of animal experimentation to elevate science to an independent status was more important than any benefits – not that he anticipated any – it might confer on humans. His frank acknowledgement of its futility was unusual since most scientists attempted to justify the practice in terms of its benefits for humans. This was not always easy; even Bernard, writing towards the end of his life, found it hard to identify cases where animal experimentation had benefited human medicine.[7] Yet the myth of medical progress had to be promoted; it was, after all, the only way of explaining the practice to a horrified and sceptical public. Although there was no evidence to support the rhetoric spouted by the International Medical Congress and the British Medical Association, the tactic of associating animal experiments with medical progress was successful. The two became inextricably linked, and objections about cruelty were countered with arguments that the needs of humans took precedence. Then, as now, the hypothetical benefits were far off in the future, but their promise served to quieten opponents.

After the 1881 Congress, scientists only grew bolder in their defence of vivisection. A year later, the Association for the Advancement of Medicine by Research was established, its apparent goal being the repeal or modification of the Cruelty to Animals Act.[14] The Association, already the recipient of the Congress's profits, had by May 1882 received over £1,000 in subscriptions. Nevertheless, its treasurer, Samuel Wilks, cautioned that the Association still needed 'sufficient funds in hand to meet promptly any attack from outside upon the invaluable labours of competent investigators'.[24] The phrase 'from outside' is informative, for while many doctors and scientists had formerly spoken out against vivisection, the entire community of medical and biological scientists now began to fall into line and support vivisection, at least in public.

'The new unanimity of the 1880s', writes historian Nicolaas Rupke, 'showed how much during the preceding decade the practitioners of the biomedical sciences had developed a sense of professional identity, closing ranks when outsiders demanded public accountability'.[14] Public opposition to vivisection served to unite scientists and doctors to get behind the practice. After all, they had much to lose; if successful, anti-vivisectionists could topple them from the prestigious positions they were acquiring and which were still, at the time, tenuous.[14] The pathologist and historian Alan Bates recalls that, as a student, he was taught that vivisection was indispensable for knowledge and that its opponents were enemies of science. 'To speak out was disloyalty, and medical students and young researchers (as I know from experience) went along with the culture of animal experimentation because to dissent was heresy.'[13] Even as recently as 2017, when I last worked within a medical school, there was still a sense that it was somehow treacherous to question the science of animal experimentation, indicating just how firmly closed the ranks remain.

The Brown Dog Affair

Near the Old English Garden in Battersea Park, London, the statue of a small terrier can be found on the top of a stone plinth. Often decorated with flowers and other items,[25] the dog looks uncertain; its ears are expectant, but it cowers slightly and wears a worried expression. Erected in 1985, it replaced the bronze statue of a proud-looking brown dog, placed almost eighty years earlier upon a granite memorial in a prominent position in Latchmere Recreational Ground, Battersea. The memorial, which was also a drinking fountain and had a water trough for animals, carried the following inscription:

In memory of the Brown Terrier Dog Done to Death in the Laboratories of University College in February 1903 after having endured Vivisection extending over more than Two Months and having been handed over from one Vivisector to Another Till Death came to his Release. Also in Memory of the 232 dogs Vivisected at the same place during the year 1902. Men and Women of England, how long shall these Things be?

The vivisection of the brown terrier had been witnessed by two Swedish feminists and anti-vivisection activists, Lizzy Lind af Hageby and Leisa Schartau, who had enrolled in 1902 at the London School of Medicine for Women under an arrangement that enabled them to attend lectures and demonstrations, including vivisections, at other London medical schools. In 1903, they published an account of their experiences, including a report of the repeated vivisection by William Bayliss of the brown dog during which, they stated, the physiologist had failed to use anaesthesia.[26] Since the Cruelty to Animals Act forbade animals to be repeatedly vivisected or to be vivisected without adequate anaesthesia, this led ultimately to a libel trial. Bayliss testified that the dog had received anaesthesia and, on 18 November 1903, the jury unanimously found in his favour.

The commemorative statue, erected in 1906, soon became the site of clashes between anti- and pro-vivisectionists, known as the 'Brown Dog Riots'. Medical students repeatedly protested against the statue and tried to remove it but each time were driven back by anti-vivisectionists. At the height of the protests, six constables a day were required to police the statue at a cost of £700 a year. In 1908, amidst this unrest, the Research Defence Society was established by scientists in an effort to persuade the public of the benefits of animal experimentation. By 1910, Battersea council had had enough of the controversy and removed the statue under the cover of darkness and the watchful eyes of 120 police officers, the bronze dog apparently being melted down later by a blacksmith. The current statue was commissioned decades later by anti-vivisection organisations, as a reminder that animal experiments continue and that the controversy surrounding them persists.

The legacy of the 1900s

Medical science today remains tightly in the grip of Claude Bernard and his peers.[27] Many contemporary scientists continue to assert that the best route to biomedical knowledge is via animal experiments, and the tactic of associating animal experimentation with medical progress endures into the present day, despite an ongoing dearth of evidence to support it. There remains an ingrained perception among funders, journal editors and many researchers that real science is based in the laboratory, with clinical, patient-based research being but a poor relation. Most damaging of all, contemporary animal researchers have inherited Bernard's disregard for the significance of species differences, as we shall explore in a later chapter.

Indeed, they appear to have inherited an indifference to theory as well. This may be due to a disinclination to speculate which seemed to accompany the mania for experimentation. As the English scientist and surgeon John Hunter apparently said to the physician Edward Jenner, 'Why think? Why not try the experiment?'.[28] Nor was Magendie very keen on theory, preferring to put his faith in what his experiments showed him. Even Bernard remarked, 'Magendie had only eyes and ears, but no brain when it came to experimentation.'[29] Yet experimental results and theory should go hand in hand; data on their own are not useful, they need to be interpreted by theory.[27] Indeed, theories are developed in an attempt to explain existing facts or data. The animal research paradigm has suffered as a result of its reliance on experimentation at the expense of theory, now appearing somewhat crude and unsophisticated.

None of these issues would prevent the rapid expansion of animal experimentation throughout the twentieth century. Nor would public opposition, since the tactics used to protect professional interests and pushback against opponents were dusted off and employed whenever necessary. (Indeed, they are making a comeback now, as we shall see in the final chapter.) Government statistics indicate that the number of experiments conducted in Britain increased from 250 in 1881, the first year that records were kept, to 95,000 in 1910. While the report of 250 experiments is likely to have been an understatement, since 622 experiments were reported in the 1873 *Handbook for the Physiological Laboratory* alone,[7] the leap in the number of reported experiments is astonishing, and the numbers continued to climb over subsequent decades, peaking at 5.6 million in 1971. What explains this enormous increase over the course of the last century?

2

STUCK

In 1877, the 'gentleman scientist' and secretary to the newly formed Physiological Society, George John Romanes, bought his dog Major a train ticket and travelled with him to London. There he proceeded to University College, where he delivered Major to the physiologist and founding member of the Physiological Society, Edward Schafer. Romanes, whom Schafer described as having a 'simple childlike nature', donated Major for the satisfaction of feeling that his dog had been of 'some use'.[1]

Dogs were frequently used in experiments around this time because they were relatively easy to acquire. The 1876 Cruelty to Animals Act did not extend to the procurement of animals for laboratories and in general this seems to have been a haphazard activity, with scientists taking whatever animals they were able to get hold of. The 1906 Dogs Act made it illegal for stray dogs to be offered or sold for the purposes of vivisection but this did little to allay public concern, and vehement protest from the Canine Defence League and anti-vivisection societies about the use of stray and stolen dogs in experiments endured well into the twentieth century.[2] Finding it increasingly difficult to obtain animals for their experiments, scientists were forced to search for a more reliable supply.

Creation of the 'standard' laboratory animal

In the late nineteenth and early twentieth century, the keeping and breeding of mice and other small animals, often with genetic traits that made them 'fancy' specimens for shows and competitions, became popular on both sides of the Atlantic. 'Without warning', went a piece in the 1898 edition of the *Harmsworth Monthly Pictorial Magazine*, 'the mouse fancy has sprung into general popularity, and the craze for rearing and showing the tiny creatures has assumed the proportions of an important and fashionable hobby'.[3] Before long, scientists discovered the commercial market that existed to serve animal 'fanciers' and began to place large orders. The dealers, however, operating within what was essentially a cottage industry, were frequently unable to fulfil the escalating requests from laboratories, and scientists often ended up with unhealthy or infectious animals that the fanciers did not want.[4]

Throughout the First World War and afterwards, research activity, including animal experimentation, increased because of the need to treat war injuries and the infectious diseases spread by mass migrations. A national fund for medical research of around £57,000 per year (the equivalent of about £4 million today)[5] had been created in the form of the Medical Research Committee, founded in 1913. This was funded by National Insurance following the 1911 National Insurance

Act, which included a clause allotting a portion of its revenue to research.[2] The Committee explicitly condoned animal experimentation but was becoming aware that variability in the species, strain and health of available animals was leading to inconsistent responses to testing. It therefore recommended establishing 'pure' types of animal with 'standard reactions'.[2] In 1920, the Committee became the Medical Research Council and established a National Institute for Medical Research in North London. Two years later, to ensure an adequate supply of healthy, infection-free animals, the Institute purchased fifteen hectares of land outside London and established its own farm to breed animals for medical research.

Rhodes Farm became the first establishment dedicated to the mass production of 'standardised' experimental animals in the UK and, by 1938, was supplying the Institute with 1,406 rabbits, 5,089 guinea pigs, 9,215 rats and 37,960 mice.[4] The publication in 1942 of a paper showing that Rhodes Farm guinea pigs responded better to immunisation procedures than guinea pigs from other sources led to rumblings within the scientific establishment. The *British Medical Journal* made a plea for 'standard guinea pigs' to be more widely available, highlighting the disadvantages suffered by those laboratories unable to access 'reliable' experimental animals.[4] By 1947, the Medical Research Council had established the Laboratory Animals Bureau in an effort to establish a national supply of 'standardised' animals and, within a few years, the Bureau was actively regulating the market via a voluntary scheme of accreditation.[4] Soon, the breeding of animals for laboratories became concentrated in a small number of establishments capable of mass production.

Meanwhile, across the Atlantic, former schoolteacher Abbie Lathrop was breeding ferrets, rabbits, guinea pigs, rats and mice on a farm in Massachusetts. Forced to retire from teaching due to ill health, Lathrop had begun supplying the 'fancy' animal market. She was particularly interested in mice and selected traits for 'creamy buffs', 'white English sables' and other coats that her mouse-fancying clientele considered desirable. But in 1902, she was contacted by William Ernest Castle, a scientist from Harvard University who was seeking mice for his genetics research. Lathrop was by now a skilled and experienced breeder and was soon busy fulfilling orders for Castle as well as other scientists.[6]

Castle put an undergraduate, Clarence Cook Little, in charge of his mouse colony, and in 1909, Little began inbreeding the mice for his studies on the biological and genetic nature of cancer. Lathrop had started her own investigations into the connection between certain strains of mice and the inheritance of cancer alongside Leo Loeb, another scientist she supplied, but it was Little's work that would gain the credit. In 1929, he founded the Jackson Laboratory in Maine, using mice originating from Lathrop's stock.[6] In the same year, he became director of what is now the American Cancer Society and recommended the diversion of government funds to animal research and mouse breeding centres. He promoted the wonders of Jackson Laboratory mice in publications as diverse as *Good Housekeeping* and the *American Journal of Cancer*, and by the 1930s, 'JAX mice' were being touted as the answer to cancer, the 'unknown enemy'.[7,8]

Mice were not alone in being recruited by American scientists. In the 1890s, a young Swiss immigrant, Adolf

Meyer, had begun breeding albino rats for his neurological research. He chose rats because the young are born in an undeveloped state, which made his experiments feasible. His aim was to compare the development of the rat brain with that of other animals, including humans.[9] Henry Donaldson, one of Meyer's associates and Scientific Director of the Wistar Institute of Anatomy and Biology in Philadelphia, was influenced by Meyer's enthusiasm for albino rats and established a rat colony at the Institute, specifically for scientific research. By the 1920s, the Institute had sold over 40,000 of its rats to laboratories across America. The Wistar rat rapidly became a commercial success and, in 1942, the Institute trademarked the name WISTARAT® to distinguish its strains from others that had begun to enter the market.[7] Breeding rodents for laboratories soon became a profitable enterprise; not only did rats and mice reproduce frequently, but there was also now a continuous demand.

The beginning of the 'animal model'

These selectively bred rats and mice were known as 'standards', partly because of the purity of their strains but also because of the uniform conditions in which they were kept; scientists tried to ensure consistency of diet and housing across different laboratories in an effort to make their experiments more reliable. However, the term 'standard', aided by the gradual erosion of diversity in these rodents, gradually acquired the meaning of 'standard animal',[7] implying a 'general animal' that could stand in for other animals. As the preface to a 1933 handbook on white rats declared, 'it is possible to write an essentially complete outline of the science of animal behavior without going beyond the available data on the rat'.[10]

In a lecture delivered to the National Institutes of Health in 2016, Professor Todd Preuss, neuroscientist and anthropologist at Yerkes National Primate Research Center, suggested that this reimagining of the rat came about partly because of the sorts of claims made by those breeding and selling them, namely, that they could be used for almost any physiological or behavioural research, or that almost any disease could be cured by studying them.[11] The notion of a 'generic' animal, an animal that can stand in for any other, including human animals, underlies the current use of animals as 'models'. The belief is that an animal can model a human system or disease sufficiently well for the results of experiments on that animal to be relevant to humans.[12] The use of animals as models took off in the post-war years, with numerous attempts to replicate human diseases in animals and then test experimental drugs on them. A search of the term 'animal model' in PubMed, the free database of biomedical research references, reveals that in 1950 the term was found in the scientific literature eighty-four times. Fifty years later, it would be identified 13,222 times, and by 2021, 48,899 times. These 'laboratory animals', neither wild nor domesticated, were a new kind of animal. Constructed as tools for laboratories, they gradually came to be regarded less as animals and more as just another piece of laboratory equipment.[13] Frequently referred to simply as 'models' in the scientific literature, it became possible to read a scientific report of animal research in which the animal itself had disappeared.

Animal use increases post-war

With a reliable supply of animals now established, the numbers of experiments climbed steadily after the Second World War, facilitated by the increasingly strong role of governments in funding biomedical research.[14] Following the bombing of Nagasaki and Hiroshima, for example, the Medical Research Council funded a large programme of research that involved subjecting numerous animals to low-dose radiation, the aim being to understand the effects of exposure on humans.[15] Further experiments began after the war at the UK government's military research facility, Porton Down, where thousands of animals were used to test the effects of chemical and biological weapons. And of course, the expansion of pharmaceutical research in the 1950s and '60s led to the most extensive ongoing use of animals, with scientists analysing their physiological responses, tissues, body fluids and behaviour to throw light on the effects of potential new drugs. Then something happened that made certain experiments a legal requirement, intensifying animal use still further.

The shocking discovery that German doctors had conducted brutal experiments on prisoners during the Second World War had led to guidelines stipulating that human research should be preceded by experiments on animals. These guidelines, established at Nuremberg in 1947, were ethical rather than legal, but then disaster struck. The prescription of thalidomide for 'morning sickness' in the late 1950s and early 1960s resulted in thousands of children being born with serious birth defects, including congenital heart disease, disorders of the eyes and ears and shortening or absence of limbs, with thousands more dying around the time of birth. As a result, the Nuremberg guidelines solidified into a legal requirement that animals be used to test the safety of chemicals, including drugs. The UK Medicines Act, introduced in 1968 as a result of the thalidomide disaster, made it mandatory to test the safety of drugs in animals before testing them in human trials. The UK also established the Committee on Safety of Drugs, known now as the Commission on Human Medicines, and many other countries responded by introducing tougher rules for the testing and licensing of drugs. With the use of animals now legally mandated for safety testing, animal research became increasingly routine and normalised.

As an interesting aside, thalidomide had been extensively tested in animals, but whether it had been tested in *pregnant* animals is unclear. After the disaster, many concluded that it could not have been, assuming that this would have averted the tragedy. Consequently, the drug was then tested in multiple species, breeds and strains of pregnant animals, but all to no avail.[16] It was eventually discovered that New Zealand white rabbits[17] and some species of monkey also delivered offspring with birth defects as long as enormous doses of the drug, much higher than those given to the pregnant women, were used. It is not surprising, then, that animal tests failed to raise safety issues, because even if pregnant animals had been used, scientists would have been unlikely to use such high doses of thalidomide or select the rare species that produce birth defects after exposure.

Animal research becomes locked in

As animals were increasingly bred to be 'pathogen free', once they arrived at laboratories they had to be housed in tightly controlled environments where they could be kept free of infection. Consequently, by the middle of the twentieth century, many large institutions began to build special housing for animals, as well as stockrooms to house breeding colonies and facilities for surgery, animal feed and equipment. Companies supplying scientists with the necessary equipment, such as surgical devices, cages and cage washers, started to emerge. Careers and livelihoods began to depend upon using animals in research, not just the livelihoods of scientists and those who supplied them with animals and equipment, but also of those within the pharmaceutical industry, private testing companies, government agencies and corporations that used products tested on animals.[18]

Within academia, various mechanisms began to lock scientists into conducting animal research. Entire scientific communities gradually built up around rats, mice and other animals commonly used in laboratories, and as the twentieth century progressed, positive feedback loops started to emerge. Scientists using Wistar rats, for example, accumulated knowledge and expertise on how to look after these rats and use them in experiments. They published their research, which put them in a good position to seek funding to conduct more research on Wistar rats. Growing familiarity with the species made it tempting for other scientists to reach for a Wistar rat rather than a different animal or method, as they were able to draw on previous scientists' experience as well as a range of 'off the shelf' methodologies for their experiments.[11] As a result, research with Wistar rats could be conducted quickly and easily, and the track record of research on the species made it easier for other scientists to gain funding for research using these rats. After a while, journals and funding agencies became so habituated to certain 'infrastructures of knowledge',[19] that they queried research that *didn't* use the usual methods or animals. Economist Dr Joshua Frank has explored how such lock-in mechanisms operate within animal research and observes, for example, that in using animals to conduct toxicity tests, a knowledge base develops that makes animal research an easier option for scientists conducting future toxicity tests, even though other methods are available.[18]

Another kind of infrastructure also began to emerge. Within academia, numerous conferences, journals, professional associations, funding organisations and academic programmes devoted to animal research became established, all of which made using animals the default option.[18] Furthermore, it soon became clear that animal research could be relatively easily and quickly converted into data and then into publications, the key currency of academia. After spending some time observing scientists at work in a biology laboratory, the French philosopher Bruno Latour was struck 'by the way in which many aspects of laboratory practice could be ordered by ... the transformation of rats and chemicals into paper'. He continued, 'Bleeding and screaming rats are quickly dispatched. What is extracted from them is a tiny set of figures. This extraction ... is all that counts.'[20]

Not all academic publications have scientific worth, but all have a professional and institutional value; the more publications, the greater one's career prospects and the better the university's reputation. By transforming quickly into publications – much more quickly than research with humans – animal research soon began to provide an efficient route to obtaining a PhD and a scientific career. Animal research publications were used to support applications for funding further animal research. Where these were successful, the funding brought wealth and prestige to universities, enabling them to attract and train up more scientists to conduct animal research.

Academics who conform to the norms of their discipline are more likely to gain funding for their research, publish their findings more easily and attain powerful positions in the field.[18] This puts them in a position to perpetuate existing norms and belief systems, locking the paradigm into place. In the 1960s and '70s, John Gluck was a successful primate researcher with his own laboratory. But unlike most scientists who, after remaining in the field of animal research for any length of time, become fully signed up to the paradigm, Gluck began to harbour doubts.[21] He recalls, among the many things that gave him pause for thought, being given a copy of Peter Singer's *Animal Liberation*[22] and, on another occasion, arriving at work to find that protesters had released some of the monkeys from his laboratory.[23]

After a period of uncertainty and moral reflection, he left the world of primate research, no longer feeling able to justify his work. This response was unusual; most animal researchers appear to double down when challenged, leading to even greater entrenchment.[18] When the modern animal rights movement began to mobilise, motivated in the UK by increasing awareness of the scale of animal experimentation[24] and the failure of the 1986 Animals (Scientific Procedures) Act to curb the practice, most scientists did not pause to reflect as John Gluck did; they simply made their laboratories more secure and retreated behind the locked doors.

As a result, the animal research community became even more isolated and impenetrable than before. Scientists using animals had little or no engagement with other disciplines or scientific ideas; they conducted research and published papers for other members of their community, all of whom shared their values. Lacking challenge from other academics and insulated from the public, their beliefs and practices ossified. Lobbying groups such as the Research Defence Society and the Coalition for Medical Progress became more vociferous in their defence of animal use, merging in 2008 to form Understanding Animal Research, which continues to defend and promote the interests of organisations that conduct animal experiments.

Animal research today

In 1974, American embryologist Beatrice Mintz created the first genetically modified animal by inserting a DNA virus into an early stage mouse embryo. By 1981, other research teams had developed techniques to enable mice to pass the 'transgene' to their offspring, spawning a plethora of studies, mostly in mice and rats. Today, genetically modified animals are used to study the function of specific genes and their relationship to health and disease. Some animals have one or more genes inactivated, in the hope that this will provide information about the roles of those genes, while others have human genes inserted in the hope of making them more human relevant.[25] The number of UK animal experiments had begun to decline after peaking at 5.6 million in 1971, falling to a low of 2.6 million in 2001, but the breeding of genetically modified animals led to a huge resurgence, causing the number of experiments to reach 4.14 million in 2015. In 2021, the most recent year for which figures are available, the 'creation' and breeding of genetically modified animals accounted for forty-three per cent of the 3.06 million scientific procedures conducted on animals in the UK.[26]

The remaining fifty-seven per cent was for three different types of research: 'basic research' intended to increase our understanding of how complex organisms function and develop; 'applied research', where animals are used to identify potential drugs and provide information about the effects of drugs that seem promising; and 'regulatory research', conducted to satisfy legal requirements for safety testing and typically involving toxicity studies to investigate the potential adverse effects of substances.[26]

Big business

A 2022 report indicates that the global animal testing market is expected to grow from $10.74 billion in 2019 to $13.8 billion in 2025 and $17.6 billion in 2035,[27] due to an ongoing demand for animal studies and advances in genetic engineering. With such projections the amounts tend to vary, but the consistent factor is that growth is anticipated. Meanwhile, an analysis of gross domestic product by a country's estimated laboratory animal use found an almost perfect correlation.[28] The philosophers Hugh LaFollette and Niall Shanks were not far off when, a quarter of a century ago, they declared that animal research laboratories were supported by a 'vivisectionist version of the military-industrial complex'.[12]

In February 2023, the net worth of Charles River Laboratories alone was estimated at $12.74 billion.[29] A huge American company, Charles River specialises in providing services for the pharmaceutical and biotechnology industries. It offers a vast catalogue of animals, including the option of a custom-made transgenic model. Among their many rodent models is the Wistar rat, still one of the most popular species used in research.[30] Meanwhile, the Jackson Laboratory, now the world's leading supplier for laboratory mice,[6] had a revenue of $551 million in 2021 (an increase of eighty-three per cent since 2015).[31] It offers mice specifically bred for different purposes, including for research into hypertension, Covid-19, Alzheimer's disease, type 2 diabetes and cancer, or as immunodeficient, 'humanised' or transgenic. Many of the companies providing equipment for laboratories also make substantial sums, with assets increasing year on year. These companies sell a range of products to scientists such as stereotaxic devices for holding animals still during experiments, 'rodent pinchers' for measuring and testing pain and apparatus designed to produce specific injuries, for example the 'spinal compression device' which applies a measurable amount of weight to the spine, to replicate spinal cord injury.

A paradigm entrenched

Over the course of the twentieth century, then, various events, developments and mechanisms combined to ensure that the animal model paradigm became embedded within biomedical research. But these forces are not immediately obvious to the casual observer, many of whom might assume that the steady escalation in the number of animal experiments, from a few thousand in the early 1900s to almost six million by the 1970s,[32] was due to their success in generating life-saving treatments for humans. Lobbying organisations certainly made strong claims about the role of animal experiments in medical developments, contributing to this view.[33] And once animal research became established, its ubiquity led to a presumption that any medical advances or health improvements were due to the practice; it was common, for instance, to attribute declining mortality rates and higher life expectancy to the rise of laboratory-based medicine.[34]

Yet, as statisticians frequently caution us, association is not the same as causation, and in the following section I explore this issue in depth. I ask whether animal research really does drive medical innovation and generate treatments for humans and I investigate the extent to which it actually does ensure the safety of new medicines. The central question here is whether the findings generated from animal research can be translated to humans. Answering this question is not straightforward, because translation is impacted by both species differences and scientific rigour. For this reason, I begin the next section by investigating first, the extent to which scientists conduct animal research according to accepted scientific standards, and second, how seriously they consider the implications of species differences. Are these critical issues given sufficient attention? If not, then the findings from millions upon millions of animal experiments will be worthless.

PART TWO

IN CAPTIVITY

3

BIAS AT THE BENCH

In 2002, I was working as a researcher within the School of Social and Community Medicine at the University of Bristol. Some researchers were using a relatively new methodology known as 'systematic review' to assess, synthesise and analyse large bodies of data – usually from human trials – in an exacting and transparent way. Systematic reviews were soon acknowledged to be an excellent way of generating robust evidence for making objective decisions relating to healthcare, and I wondered about their potential for the field of animal research.

I discussed the issue with epidemiologist and Head of School at the time, Professor Shah Ebrahim, with whom I had worked previously. We got together with Professors Michael Bracken, Ian Roberts and Peter Sandercock, also epidemiologists, who were aware of the need for more robust evidence in this field[1] and decided to investigate further. We found that few systematic reviews of animal data had been conducted and concluded that only flimsy evidence existed to support the use of animals in research intended to have relevance for humans. We wrote a paper to this effect, arguing that more systematic reviews of animal studies should be conducted.

Since animal research was particularly controversial at the time, with many laboratories the focus of regular anti-vivisection protests, Shah Ebrahim gave advance notice of the paper's publication to the Dean of Medicine and the Director of Research at the University of Bristol. Out of courtesy, he also discussed it with Colin Blakemore who was then Chief Executive of the Medical Research Council, since the School was in receipt of Medical Research Council funding. All of those consulted strongly advised against publishing the paper, declaring that it would provide ammunition for the anti-vivisection movement and – astonishingly – stating that it constituted an attack on essential biomedical research. We were also told that we were not in a position to understand animal experiments. Disturbingly, Shah Ebrahim was told that publishing the paper would not enhance his career prospects.[2]

The paper,[3] published in the *British Medical Journal* and reported in the national media, met with hostility from the scientific establishment, as noted in the Preface. Nevertheless, other responses indicated that the paper – and the debate it provoked – was welcome. What is certain is that over subsequent years, the number of systematic reviews of animal studies increased. Furthermore, two centres were established to promote and support the practice of systematically reviewing animal data: CAMARADES (Collaborative Approach to Meta-Analysis and Review of Animal Data from Experimental Studies) in the UK and SYRCLE (Systematic

Review Centre for Laboratory Animal Experimentation) in the Netherlands, in 2004 and 2012 respectively.

Animal studies were now under scrutiny in a way that they had never been previously. Shockingly, as a result of researchers all over the world poring over hundreds and then thousands of these studies, it soon became apparent that the scientists using animals had not been doing a very good job. The quality of animal research was found to be substandard, with poor adherence to even the most basic scientific principles. Of course, scientists using animals are not the only ones to conduct poor quality research; indeed, researchers working with humans have been struggling with many of the issues we're about to explore for decades.[4] What sets animal researchers apart, however, is the deeply controversial nature of their work. Two-thirds of people in the UK report that the use of animals in scientific research 'bothers' them, and such support that does exist is highly conditional.[5] Few would be likely to support this research if they knew it was being conducted with such little regard for scientific rigour that most of the findings could not be trusted and that countless animal lives were being wasted as a result. Yet unfortunately this is the case and we, the public, are paying for it. So, what is going wrong? What have systematic reviews and other analyses of animal research revealed?

Bias

One of the most important challenges in any kind of research is to minimise the risk of bias. Researchers usually have a certain view of how they want a study to turn out and – without necessarily being aware of it – are able to influence its outcome unless measures are taken to guard against this. Any sort of bias introduces doubt about whether we can trust a study's findings.

In a typical drug test using animals, the experimental drug is administered to a group of the animals whose responses are then monitored. To compare animals that have had the experimental drug against those that have not, there should always be a control group. Animals in control groups receive an inert substance (sometimes called a placebo) rather than the experimental drug. However, unless animals are allocated to the different groups randomly, researchers might, for example, choose the stronger looking animals to go into the group having the experimental drug. If that group then has better outcomes than the control group, it might be concluded that this is due to the experimental drug, when in fact it could be due to the fact that the animals were already stronger and healthier. Another manifestation of this sort of bias occurs when researchers give those animals receiving the actual drug more positive scores when assessing the outcomes of that treatment. For these reasons, the scientific community has agreed that specific steps need to be taken to avoid these biases, which are often unconscious. These steps include random allocation, a method for randomly assigning animals to treatment or control groups; allocation concealment, a way of concealing this allocation from those assigning animals to treatment or control groups; and blinded outcome assessment, a way of concealing from those assessing the outcome of experiments which group the animals belong to.

Unfortunately, awareness of these potential sources of bias is low amongst animal researchers.[6] Between only twelve per cent[7] and thirty-seven per cent[8] of animal studies report using random allocation, between zero[8] and fifteen per cent[9] report using adequate allocation concealment, and between zero[8] and thirty-five per cent[9] report using blinded outcome assessment. This means that, at the very best, an astonishing two-thirds of animal studies do not take appropriate measures to minimise bias, rendering their findings unreliable.

I first met Professor Merel Ritskes-Hoitinga in the summer of 2019, when she was Professor of Evidence-Based Laboratory Animal Science at Radboud University in the Netherlands. A friendly, down-to-earth woman, she put me at ease immediately, telling me about her work to promote the use of systematic reviews in animal research and about SYRCLE, the centre she had established to provide guidance and support for scientists conducting these reviews.

'It is a very puzzling situation at the moment, that our science education is so poor,' she said. 'Many do not know what randomisation and blinding is. It is all taken for granted that science should be of high quality, but it is not. This is very worrying, because we use animal studies for human health, so we conclude all sorts of things that are not based on evidence.'

There is also a risk of bias if researchers fail to report how they calculated the number of animals to include in their study. A sample size calculation provides reassurance that a study includes sufficient animals at the outset to answer the question posed. Yet only one per cent of animal studies report how their sample size was calculated.[10] Statisticians are rarely consulted by animal researchers for help with setting appropriate sample sizes and even senior animal researchers report that they generally choose sample sizes based on what has 'worked' in previous experiments, or on other non-scientific factors such as the number of cages available.[11] Unsurprisingly, sample sizes in animal research are frequently inappropriate, typically being too small to answer the research question posed. This means that animals' lives are wasted because the study cannot contribute useful information.

Confounded

Sean Scott learned about amyotrophic lateral sclerosis (ALS), also known as motor neuron disease, after his mother was diagnosed with the disease in 2001. Soon after, he became president of the ALS Therapy Development Institute, the world's largest ALS research centre. Observing that while many drugs for ALS were found to extend the lifespan of mice, but only one showed any efficacy in humans, he decided to investigate.[12]

In any research study there are factors that can confuse, or 'confound', the results unless they are properly managed. For example, if extended lifespan is the measure of a drug's effectiveness, then it is necessary to ascertain whether there are any other factors that might also extend lifespan. If there are, these potentially confounding factors should be taken into account when designing the study or analysing its results. In animal research, the range of potential confounding factors is huge and can include the animal's sex, age, weight and health status, laboratory conditions,[13] including diet and housing, the environment in which it was reared,[14] how it is handled, the method of administering the experimental drug, any pain or stress it might be experiencing and so on.[15] Indeed, the

University of Bergen advises the staff in its animal facility not to wear perfume or aftershave because perfumes often contain chemical neurotransmitters that can disrupt reproduction and cause stress, impacting test results.[16]

Scott retested several of the ALS drugs, this time correctly taking into account the potential confounding factors, and found that none of the drugs now extended the mice's lifespan. He concluded that most of the previous 'positive' findings had probably been due to confounding factors rather than the actual effects of the experimental drugs. Unfortunately, poorly conducted studies do not translate into effective treatments for humans. Scott himself was diagnosed with ALS in 2008 and, in the absence of any effective treatment for his disease, died a year later at the age of thirty-nine, shortly after the publication of his important paper.

Too good to be true

Studies that fail to reduce bias and deal with confounding factors are more likely to report 'positive' findings; in other words, they are more likely to conclude that an experimental drug is effective than studies that *do* reduce bias and manage confounding factors. For example, it is estimated that less than half of the published animal studies of stroke drugs that show a positive outcome are in fact truly positive,[17] and that this is due to bias.[18] Elsewhere, researchers tracked the progress of a drug for skin cancer as it went through the drug development process. As it progressed from animal to human trials, they found that the scientific rigour of the research increased and, concurrently, the apparent effectiveness of the drug diminished.[19] By 2012, Stanford University epidemiologist Professor John Ioannidis concluded that consistent evidence of serious biases in animal studies made it 'nearly impossible to rely on most animal data to predict whether or not an intervention will have a favourable clinical benefit-risk ratio in human subjects'.[20]

Mimicking the human situation

In addition to properly designing their studies and addressing all potential biases and confounding factors, animal researchers have to 'model' or simulate the human disease of interest, mimic the context in which the drug will be administered to humans and represent the relevant human population as far as possible if their research is to be valid.

The term 'animal model' is used to refer to the manipulation of an animal such that it mimics the human disease being studied. To give an example, an animal model of human obesity might be created by feeding rats a particular diet, or by operating on them, or genetically manipulating them to render them obese. They might then be given an experimental drug to see if it helps them lose weight. The key issue is how well an animal model mimics the human condition under investigation. Of course, the closer the model to the thing being modelled, the better. As scientists Arturo Rosenblueth and Norbert Wiener put it, 'the best material model for a cat is another, or preferably the same, cat'.[21]

Although there have been some successes with diseases that are based on single gene defects, there is growing recognition that most animal models lack the sophistication necessary to accurately mimic human diseases. It's also the case that we often don't know enough about the human disease in question to develop a model that mimics it.[22]

H Shaw Warren is an infectious disease specialist at Massachusetts General Hospital in Boston and Associate Professor of Pediatrics at Harvard Medical School. He is interested in inflammation, a process that is an essential response to harmful stimuli but which can sometimes get out of hand and cause the immune system to overreact, leading to a life-threatening emergency known as sepsis. A commonly used mouse model involves injecting an endotoxin (a toxin present inside a bacterial cell that is released when it disintegrates) to induce inflammation and then testing drugs to block the inflammatory response. However, although many experimental drugs were found to work in these mouse models, they failed every time they were tested in critically ill humans.

Warren wondered whether the mouse models were actually mimicking what happens in humans with sepsis. In 2003, he took a year's sabbatical at the Pasteur Institute in Paris and read everything he could lay his hands on about research that had been conducted in animals. He was shocked to find that mammals differ hugely in their response to endotoxins and, in particular, that mice are between 1,000 and 100,000 times more resistant to endotoxins than humans. Yet, for an animal model to accurately reflect the human situation, he notes, the animal used would ideally have the same sensitivity to endotoxins as humans.

On returning to Boston, Warren took part in a collaboration to study the gene responses of patients who had inflammation due to injuries and burns, as well as those who volunteered to be injected with a tiny amount of endotoxin. The researchers also studied the gene responses in mice who were either burned, injured or injected with the same endotoxin as humans. The findings were controversial and shook the world of animal research.[23] Essentially, they revealed that the correlation between the gene responses in mice and humans was 'almost random'. In a 2013 TED Talk about this research, Warren reflects that this was not altogether surprising, since mice have lived for millions of years in environments teeming with microbes.

'It seems like mice have evolved a different, and maybe even a better strategy of dealing with infection, by tolerating larger doses of microbes without inducing the same overwhelming inflammation that we see in people,' he said, before concluding, 'extrapolation from mouse models all the way to a complicated, human inflammatory disease, might be overreaching.'[24] Dr Mitchell Fink, then Professor of Surgery at the University of California, was shocked at the findings: 'When I read the paper, I was stunned by just how bad the mouse data are,' he said. 'It's really amazing – no correlation at all.'[25]

It is this inability to reproduce the complexity of human disease that is the problem. A human stroke is quite different from a stroke artificially induced in an animal. We might suffer a stroke after many years of smoking, drinking alcohol and eating unhealthily, resulting in extra weight and high blood pressure. But animal models of stroke commonly involve blocking the middle cerebral artery, which represents only the final stage of a process that in humans is many years in the making. Most human diseases, and especially chronic diseases, tend to evolve over time and animal models are unable to mimic their slow, progressive and degenerative nature. Rheumatoid arthritis, a chronic, systemic, autoimmune disease affecting the connective tissue, provides another example. While characterised by joint inflammation and bone and cartilage erosion, the disease can also affect the skin, heart, lungs and kidneys. Many animal models of rheumatoid arthritis exist, but while they may exhibit some of its features (e.g. joint swelling), none is able to capture the complex functioning of the human immune system. Many genetic variations affect how an individual immune system performs and these, in combination with environmental factors, determine whether an autoimmune disease such as rheumatoid arthritis will develop. Consequently, while animal models may have been able to elucidate some of the mechanisms of rheumatoid arthritis, they are not suitable for developing treatments.[26]

It is not surprising that animal models are unable to capture the complexity of human disease, but what *is* astonishing is the ability of some scientists to deny this complexity. Cancer specialist Professor Azra Raza, of Columbia University's Herbert Irving Comprehensive Cancer Center, finds this incomprehensible. In her best-selling book, *The First Cell*, she describes cancer as an 'evolving, shifting, moving target, far too impenetrable to be deconstructed systematically, far too dense to lend itself in all its plurality to recapitulation in lab dishes or animals'.[27] And, as she notes, although cancer may appear to strike suddenly, it is actually the result of a gradual accumulation of small changes, intricately tied to ageing.

Some animal models of cancer involve culturing human tumour samples as cell lines and implanting these into mice. The immune systems of these mice have to be destroyed first so that they don't reject the human cells. However, as Raza points out, this means that the mice cannot possibly represent the microenvironment of the human body in which cancer cells thrive. Moreover, she notes that cells change characteristics as they adapt to new environments, meaning that once inside a mouse, the cells will not be exactly the same as those in the human tumour.[27] Having worked with cancer patients for over thirty years, Raza knows that every person's cancer is unique. Consequently, she struggles to understand why some scientists believe that animal models can successfully represent it.

Animal models can appear remarkably crude, and perhaps none more so than when they are applied to human mental health. One animal model reduces the complexity of human depression to a rodent's loss of preference for sweet fluids following exposure to a prolonged series of stressors.[28] Scientists have even suggested that animals be used to investigate the mental health effects of social isolation due to Covid-19 lockdowns.[29]

The importance of context

A further condition that has to be met if the findings from an animal study are to be generalisable to humans is similarity of context. In other words, the circumstances in which animals are given an experimental drug need to be similar to those in which humans will be given the drug. There was great excitement about the experimental drug tirilazad after it was found to be effective when given to animals ten minutes after having a stroke induced. In the human context however, it is impossible for a patient to get to hospital, have the appropriate tests, receive a diagnosis and be given the drug within ten minutes of stroke onset. In human trials, the drug was administered within a more realistic five hours of stroke onset and found to be unsuccessful.[30]

In the field of multiple sclerosis, experimental drugs are most commonly administered to animals some days *before* neurological impairment is induced. As these drugs may work by blocking the induction of the disease, they are not relevant to the human condition because there is no way of identifying human patients prior to the onset of the disease. Animal models of multiple sclerosis will only be relevant to the human condition if the treatment can be successfully started *after* the onset of symptoms, not before.[31] These are not selected examples; the timing of experimental interventions in animals has dubious relevance to the human situation in many fields, including Parkinson's disease,[32] inflammatory bowel disease,[33] rheumatoid arthritis[26] and osteoarthritis.[34]

Furthermore, the doses of drugs used in animal studies often bear little relation to those used in the first human trials, and most animal studies administer drugs differently from the way in which they will be administered in humans.[35] In the context of rheumatoid arthritis, Dr Cathalijn Leenaars, from Hannover Medical School in Germany, compared animal studies of the drug methotrexate with the corresponding human trials and found that the dosing schedules and routes of administration were very different. A large proportion of animals had the drug administered intra-muscularly, for example, while for humans the most common route of administration was by mouth.[36] All these discrepancies have implications when extrapolating from animal studies to humans.

Representing human populations

Yet another necessary condition for extrapolating from animals to humans is that the animals represent the human population in question as closely as possible. For example, if animals are being used to investigate treatments for human stroke, then the animals studied should be elderly, like the human stroke population. The animals used in stroke research, however, are usually young and healthy.[37] As a result, animal studies frequently overestimate the benefits of experimental stroke drugs. Once the drug is tested in elderly stroke patients, who may have pre-existing conditions such as high blood pressure or diabetes, and who may be taking a range of medications, the drug turns out to be not as effective as the studies in young healthy animals suggested.[17] There are many such examples where the disparity between animal and human study populations hampers generalisability.

We have also seen that laboratory animal populations are highly standardised, meaning that the animals are genetically very similar to each other. The rationale behind this is to make everything as uniform as possible, but this uniformity makes it difficult to extrapolate the results to naturally diverse human populations. Being from such highly inbred strains, most laboratory animals cannot even be reliable models for their own species, let alone another.[22] 'Imagine you were doing a human drug trial', writes Joseph Garner, Associate Professor in the Department of Comparative Medicine, Stanford University, 'and you said to the FDA [Food and Drug Administration], "OK, I'm going to do this trial in 43-year-old white males in one small town in California ... and they all have the same grandfather!" ... that's exactly what we do in animals. We try to control everything we can possibly think of, and as a result we learn absolutely nothing.'[38]

Because of the homogeneity of laboratory animal populations, individual variation in disease susceptibility, immune response and drug reactions was for decades a neglected area of study, with grave repercussions for human medicine.

Concealing troublesome findings

So far, we've considered the many challenges involved in conducting a scientifically rigorous animal study, but the story doesn't end there. Problems also arise when the research findings are analysed and reported. Basic information, such as the exclusion of individual animals from experiments, is often very poorly described,[39] yet such exclusions can dramatically alter the results of a study. For example, an animal might be excluded from a study if it becomes ill or dies, but if it is excluded from the study and the reader is not told about this, then the experiment will appear more positive than it actually is.[40]

Negative findings – those that don't support the researcher's hypothesis – are of course just as important as positive findings, but unfortunately the latter are often considered more exciting and newsworthy. Because of this, researchers whose studies don't at first produce positive results may be tempted to try out different analyses until they manage to unearth some.[41] They may then report only these positive results in publications, possibly also altering their original research question to align with the new findings. Such practices are not only unscientific, they also lead to a whole body of research with an inflated proportion of positive studies.

Epidemiologist Konstantinos Tsilidis, from Imperial College London, used a statistical technique to estimate whether the number of published animal studies with positive results was too large to be true. Looking at over 4,000 animal studies, he and his colleagues calculated that only 919 of these might be expected to have positive results. In fact, 1,719 reported positive results, leading Tsilidis and his colleagues to conclude that the excess positive studies were due to the scientists having been selective in the analyses and findings they chose to report.[42] The way to overcome this is for researchers to publish their research protocols in advance, specifying up front what their questions are and what analyses they will perform, so these cannot be changed as they go along. This is an important issue because over-optimistic animal study findings mislead; human trials may be launched as a result of 'positive' animal studies but, as we will see, these often end up doing little more than creating false hope, exposing humans to unnecessary risk and wasting scarce research resources.[43]

Failure to publish negative or inconclusive findings is another consequence of the emphasis on positive results.[44] Indeed, around a third of animal studies never lead to even one publication.[45,46] This means that the studies that *are* published are biased in favour of positive findings. Publication bias is a significant problem in animal research because, once more, it can make whole fields of animal research appear more successful than they actually are. In the field of stroke, for instance, the benefits of animal studies are estimated to be overstated by as much as one-third due to negative studies not being published,[47] while another study found the benefits of a cancer drug to be inflated by as much as forty-five per cent for the same reason.[8] The non-publication of negative studies also results in the loss of valuable data; negative findings are just as important as positive findings, yet if these findings do not see the light of day, other scientists cannot learn from them.

The non-reporting of animal studies is also, of course, deeply unethical. Dr Mira van der Naald and colleagues, from the University Medical Center in Utrecht, estimated that of 5,590 animals used for research within their centre, only twenty-six per cent were reported in resulting publications.[46] If the data derived from animal experiments are not published, the contribution of the animals involved in those experiments is lost, meaning that the moral justification for using these animals is also lacking.[48] Moreover, other scientists may conduct similar experiments using further animals because they are unaware that the 'failed' experiments have already been performed.

Should we try to improve animal studies?

The possibility of bias lurks at each and every stage of animal research. Each bias, if not addressed, ultimately has the effect of making animal research appear more successful than it actually is. This, in turn, creates an illusion among scientists, funders, politicians, patients and members of the public that animal research 'works'. This illusory success leads to more funding and so it continues, yet all of this conceals the fact that animal research is failing at every turn. Poorly conducted studies do not produce useful information and are a waste of money. So, should our efforts go into improving animal studies in an attempt to make them more valid and reliable?

Multiple initiatives have been launched with this goal in mind.[49] Courses are available to help scientists gain more awareness of the risks of bias and to design their studies appropriately. Tools have been developed to help researchers assess both the strength of evidence coming from animal studies and the 'translatability' of animal models. Those conducting animal research are encouraged to register their studies in advance so that each study's hypotheses and methods are on record, the hope being that this will encourage researchers to stick to their protocols and report their findings appropriately. There are suggestions that all animal study data should be deposited in publicly accessible databases so they can be shared. Some journal editors request that authors adhere to certain guidelines when submitting their research for publication.

Among the recommendations are some with disturbing implications. One is a suggestion that additional diseases are added into animal models, to make them more closely resemble situations where patients have more than one health condition. For example, it has been suggested that obesity could be incorporated into animal models of Covid-19, or that more severe versions of this disease could be induced in animals (leading, for example, to multi-organ failure), so that the human situation can be more faithfully represented.[50] Elsewhere, researchers have proposed that using larger animals, such as non-human primates, would improve the likelihood of animal studies translating to humans,[51] despite the fact that studies using large animals suffer from the same limitations as those using small animals.[52] Attempts to modify animal studies in these ways would increase the extent and severity of animal suffering, but would they increase the likelihood of the studies translating to humans?

'I have seen for thirty years that we have been trying to improve the quality of animal studies and publications, but we hardly see improvements,' said Merel Ritskes-Hoitinga, now Professor in Evidence-Based Transition to Animal-Free Innovations at Utrecht University, in a 2022 panel discussion. 'There are also examples where quality has been improved a little, and we don't see a better translation. So, I think it's better to put our efforts on new approach methods and new approach thinking.'[53]

A study she was involved in examined the quality of animal research in 2009, and then again in 2018. Needless to say, the intervening nine years brought only modest improvements.[54] In 1999, recommendations and guidelines for improving the standard of animal research into stroke were presented with great fanfare, but almost a quarter of a century later, they have not borne fruit;[55] animal studies conducted according to the best available guidelines have failed to translate into effective treatments for human stroke patients.[56] So even if the next several decades were spent attempting to resolve all the quality issues in animal research, the end result is likely to be unimpressive. At best, the quality of animal studies and their relevance to humans might improve *a little*.[57] More importantly, however, these quality issues are trumped by a much larger and more fundamental challenge to the paradigm, the problem of species differences.

4

ELEPHANT IN THE LAB

In November 2019, I travelled to the Netherlands to attend a symposium at which Merel Ritskes-Hoitinga had invited me to speak. In a fine old university building in the centre of Utrecht we – a diverse group of researchers from academia and industry, as well as scientists using laboratory animals, students, bioethicists, philosophers, veterinarians, statisticians and epidemiologists – gathered to discuss the translation of animal research to human medicine. I spoke about a study I had just completed with my colleague Rebecca Ram, about stroke researchers' views on the failure of animal studies to translate into treatments for stroke patients. I relayed to the audience that the authors of over a quarter of the eighty papers we reviewed suggested that species differences could explain the lack of translation to humans but that, remarkably, when proposing solutions, only one of these authors recommended abandoning animal models and using human biology-based methods instead.[1]

I found it extraordinary that scientists could acknowledge

the problem of species differences yet be so disinclined to try something different. Species differences represent a fundamental challenge to animal research that is intended to have relevance for humans. It's not an issue that can be resolved, and as we'll see, genetically modifying animals to 'humanise' them is not a solution.

This really is a very large elephant in the room. Researchers appear to simply hope that animals are *sufficiently* similar to humans that they can be used to model human diseases. On the website of Understanding Animal Research, the UK organisation representing the interests of the animal research community, one of the 'frequently asked questions' is: 'Are animals too different from people for animal research to be valid?' to which the following response is given: 'Obviously there are differences between animals and people. But under the skin, the biology of humans and other animals, particularly mammals, is remarkably similar. We have the same organs, controlled by the same nerves and hormones, as many other species.'[2] This is a curiously simplistic view for an organisation representing scientists, since it is the differences 'under the skin', those that are not immediately obvious, that are most significant. Even closely related species, for example, can use different biochemical pathways to achieve the same biological end, and even *within* species, differences in strain can lead to distinct differences in biological function.[3,4]

The importance of evolutionary theory

Contemporary evolutionary biology is the study of evolutionary processes such as natural selection and speciation, in other words, the ways in which different organisms evolve and adapt to become distinct species and produce such enormous biodiversity on our planet. In their groundbreaking book, *Brute Science*, philosophers Hugh LaFollette and Niall Shanks carefully and painstakingly argue that the theory of evolution fundamentally undermines the practice of conducting research on animals to develop treatments for humans.[5] They agree with evolutionary biologist Douglas Futuyma who describes the theory as 'the central unifying concept of Biology' and 'one of the most influential concepts in Western thought'.[6]

As LaFollette and Shanks explain, biological organisms are usually built from similar parts (e.g. biochemicals, metabolic pathways), but because they are faced with different evolutionary pressures, they find different ways of organising these parts to achieve similar functional ends. So even if two species have similar biological properties, this doesn't mean that the causal mechanisms underlying them are similar. Furthermore, a change in one place in an organism often has ripple effects elsewhere in the organism. Hence, they suggest that while humans are not essentially different from rats, we are differently complex.

LaFollette and Shanks note that Darwin's theory of evolution involves recognising that while some species may in many respects be biologically similar, they are also likely to be significantly *different* biologically, because they have adapted to different ecological niches. Evolutionary biologists see members of different species as being qualitatively different from each other, but those conducting animal research, following in the footsteps of Claude Bernard, tend to see things *quantitatively*. When considering differences between species they focus on differences of degree, for example in body weight or metabolic rate, and attempt to accommodate these differences using scaling techniques. This sort of thinking allows them to believe that one animal can stand in for another. But overlooking and downplaying the *qualitative* differences and their significance for research is, according to contemporary evolutionary biologists, a serious error, because even very small differences between species can result in radically divergent responses to the same stimulus. Consequently, we should not expect phenomena observed in one system to necessarily occur in another.

Yet this is exactly what animal researchers believe. Cherry picking, they use Darwin's findings on the common ancestry of animals and humans – the principle of homology – as justification for using animals to understand and develop treatments for humans, while at the same time ignoring his emphasis on the *uniqueness* of each species, because this undermines their endeavour. Using the principle of homology, they hypothesise that since animals and humans have some similarities, what is found in an animal will be applicable to humans.

In this regard the ninety-nine per cent genetic similarity between humans and chimpanzees is frequently noted, the assumption being that because two biological systems have shared ancestry, they are likely to function similarly.[7] However, the similarity between humans and non-human primates is only superficial. The key issue is not genetic similarity but the way in which those genes are *expressed*. Gene expression is the process by which the information in a gene is made into products such as proteins, which perform a vast range of functions within living organisms. Even minor differences between humans and animals in terms of gene expression can cause significant disparities in biological processes and outcomes.[8]

The uncertainty this causes when attempting to translate findings from animals to humans is the reason why Professor of Molecular Genetics at the University of Exeter Medical School, Lorna Harries, no longer uses animals in her research.

'One gene can make a bunch of different RNAs [ribonucleic acids], which are the instructions to make a bunch of related, but subtly different, proteins,' she explained, 'but the mix of RNAs that your body makes is often not the same mix of RNAs that are made in a monkey, or a mouse, or a dog or anything else. If you take all the messages and think of them together, they probably do translate across reasonably well, but not always. And when you're looking at the nitty gritty, the fine detail, only about twenty-five to thirty per cent of those things are the same between mice and men.'

It is the nitty gritty that's important when it comes to drug discovery, as we shall see in a later chapter; small differences can have significant (and sometimes fatal) consequences when extrapolating from animals to humans. Data from animal tests are not sufficiently reliable for predicting the likelihood of specific effects in humans, particularly with regard to the enzymes involved in drug metabolism, which are very different in humans and laboratory animals.[9]

Mice and humans

In 2016, Robert Perlman, an evolutionary biologist now retired from his professorship at the University of Chicago, published a thought-provoking paper about the use of mouse models of human disease. In it, he acknowledges that humans and mice (the most frequently used animals in research) have a high level of genetic similarity due to shared ancestry, as well as many biochemical and physiological similarities.[10] He emphasises, however, that the lineages that led to modern rodents and primates diverged around eighty-five million years ago and that, since then, the species in these lineages have become adapted to very different environments. Mice and humans now have very different life histories, he notes. They eat different diets, have different levels of physical activity, are exposed to different environmental toxins and pathogens and have different microbiomes. This has led to evolved differences between human and mouse immune systems and in the relationship between genetics and disease for the two species. Critically, Perlman cautions that while animals may be useful for understanding processes that arose early in evolution and that humans share with other species, they are less likely to be useful for understanding chronic, non-communicable diseases in humans because the causes of these diseases are enmeshed in our own unique human histories and cultural environments.

'What is so frustrating,' he told me, 'is that everybody justifies their research by saying, you know, what we have found in mice may be important for humans. It may be, but maybe not.'

Taking the example of cancer, he suggests that we've learned a lot about the fundamental biological mechanisms from studying mice. 'But,' he cautions, 'studies of carcinogenesis and cancer therapy have basically been a total waste. Mice are very good for giving out basic biological mechanisms, and that knowledge must be important, but how you're going to apply that knowledge to study human disease is the challenge.'

Perlman knows from his lifelong study of evolutionary biology that the differences in the ways species have evolved limit our ability to generalise from one species to another and, in this regard, he emphasises the importance of life history theory.

'We humans have all kinds of anti-cancer defences,' he explains, 'because we've evolved to live long lives and so we've evolved to repair cellular damage and prevent aberrant cells from replicating. Mice don't have these sorts of safeguards against cancers because they only live for a few years and they don't put their energies into maintaining their bodies. So, all of these treatments that can kill cancers in mice just don't work in humans because our cells have evolved; our bodies have evolved to function differently.'

Perlman suggests that scientists don't realise the barriers that different life histories present when attempting to translate research from one species to another. When I ask him if he has ever discussed these issues with scientists who conduct animal research, he sighs.

'What has struck me the few times I've given talks on animal research to the people that actually do this research, the criticism I get is, "yes, we know these mouse models are flawed, but they're the best we have and so we continue to use them".'

Same animal, different clothes

LaFollette and Shanks cite molecular biologist, Graham Cairns-Smith, who once wrote that members of different species are simply 'the same animal dressed up differently'.[11]

This notion of a 'generic' animal can be traced back to the early twentieth century, as we learned in Chapter 2, with the breeding of 'standardised' laboratory animals such as the Wistar rat. We saw that with the increasing use of these rats, scientists began to believe – without any evidence – that rats were a sort of 'generic' mammal and that the results derived from them were generally applicable to other animals, including humans.[12] Over the decades, a handful of other species including mice, zebrafish and fruit flies also became favoured generic models for conducting scientific research. Professor Todd Preuss, neuroscientist and anthropologist at Yerkes National Primate Research Center, cautions, however, that the use of these 'elite' model species has little to do with their ability to generalise to humans.[13] His observation is supported by research which has found that the most common reasons for selecting a particular animal model are the availability of the model and expertise in using it. The ability of the model to reproduce similar disease pathology and symptoms to the human condition of interest is of tertiary importance only,[14] suggesting that animal researchers are not overly concerned about the human relevance of their models. This is not far from the approach of Bernard, who believed that the choice of animal used in research should be determined primarily by convenience.[5]

Preuss does not believe that scientists should stop studying animals altogether, rather that there should be greater critical appraisal of species differences and a more logical approach to the selection of the species used in experiments. However, in a lecture delivered to the National Institutes of Health in 2016, and drawing upon his many years of research experience, he nonetheless stresses that the animal model paradigm fosters a view that all animals are basically similar, despite considerable evidence to the contrary.[13] This emphasis on the similarities, he argues, makes it very difficult to talk honestly and openly about the differences and creates an assumption that the results from one species can be generalised to other species, including humans, without actually having to demonstrate whether or not this is possible.

'Every lineage has undergone its own independent history of adaptation and specialisation,' he emphasises in his lecture. 'Every species is special. That's why we call them species. It's the same root for the same reason.'

Preuss suggests that some scientists' refusal to acknowledge species differences means that 'we end up doing some violence to some basic concepts of biology'. He notes, for example, that some scientists claim that the visual systems of macaques and humans are identical. The notion that macaques and humans are identical in *any* respect, he submits, is difficult to swallow, given that twenty-five million years have passed since they shared a common ancestor. Similarly, he observes that in the neuroscience literature it is still common to read about the 'phylogenetic scale': a hierarchical ordering of life from the simplest to the most complex organism. Referring to the phylogenetic scale as a 'folk science notion of evolution', Preuss notes that although it was jettisoned from evolutionary biology in the 1970s and is no longer considered to be a respectable biological concept, it remains in common usage within contemporary neuroscience. Aware that attachment to outmoded concepts makes scientists vulnerable, he suggests they heed the growing public scepticism:

'The public is very cynical about scientific research — if you find something in rats, well that's just rats. Well, yeah, it *is* just rats. They're not wrong to be cynical about that.'[13]

Humanising animals

One response to the thorny issue of animal–human species differences has been to try and make animals more 'human' by inserting human genetic material into them. This does not always result in the anticipated effect, however, because genes are influenced by environmental factors, including the internal environment of the animal and the external environment in which that animal lives. So, as neurologist and public health specialist Dr Aysha Akhtar from the Center for Contemporary Sciences observes, if a human gene is expressed in mice, it is likely to function differently from the way it functions in humans because it will be affected by physiological mechanisms that differ between humans and mice.[15] Preuss puts this another way, recalling that the first thing he learned in genetics is that the function of genes depends on the biochemical context, the other genes that they're expressed with.

'We've known this for a hundred years,' he says in his lecture, 'and yet, somehow, we ignore this very basic principle in biology.'

While there has been an explosion of studies using transgenic animals, then, this has spectacularly failed to produce effective new treatments for humans.[16] Studies in transgenic mice, for example, have contributed to understanding some Alzheimer's disease pathways and resulted in hundreds of experimental interventions in animals, but none has translated into disease-modifying therapies for humans with Alzheimer's.[17] Talking to Perlman about transgenic animals, it was clear that while he was in awe of the science, he nonetheless felt it was going in the wrong direction.

'You can put human genes into mice,' he says, 'and that's all very exciting, but they're still mice.'

'Do human genes behave differently if they're inserted into mice?,' I ask him.

'Yes, their regulation will be different, and the interaction of their protein products will be different because they're in a totally different milieu. You know, it's certainly neat. You can put the gene for sickle cell haemoglobin into mice and make mice that can produce sickle cell haemoglobin. But I'm not aware that that has led to any benefit in therapy for kids with sickle cell anaemia.'

Perlman notes that researchers are increasingly studying human tissues in the laboratory, which he thinks will be much more meaningful for studying human disease.

'These are obviously challenging experiments,' he cautions, 'but I think they have the potential of being much more important.'

Treatments for patients

And this, then, is the point. Animal studies are meant to be relevant to humans; the benefits to humans in terms of medical progress and new treatments is how these experiments are justified to citizens uneasy about the morality of the practice. But evolutionary theory indicates that species differences are likely to undermine the relevance of animal studies to humans. In dismissing the significance of species differences, animal researchers are pursuing an outdated understanding of biology and hindering medical progress.[5,18] They effectively dismiss one of our best biological theories, one that is confirmed by a continuous accumulation of evidence, and this places them at odds with the other biological sciences that are united by evolutionary theory.

Over a thousand years ago, the renowned Persian physician Ibn Sina issued a strong warning about the pitfalls of using animals to develop medicines for humans. Although he relied on a humoral understanding of the human body, his point about species differences still holds.

'Experiments should be carried out on the human body. If the experiment is carried out on the bodies of (other animals) it is possible that it might fail for two reasons: the medicine might be hot compared to the human body and be cold compared to the lion's body or the horse's body ... The second reason is that the quality of the medicine might mean that it would affect the human body differently from the animal body ... These are the rules that must be observed in finding out the potency of medicines through experimentation. Take note!'.[19]

Unfortunately, we have not taken note and there have been consequences. In the following chapters we explore these, considering the extent to which animal research has led to treatments for humans and ensured the safety of new drugs. First, however, we investigate the extent to which clinicians and medical researchers actually draw upon the findings generated by animal studies.

5

VITAL AND INDISPENSABLE?

As a young child in Taiwan, Frances Cheng was interested in animals and had a natural curiosity about them. Her science curriculum involved the dissection of animals, starting with earthworms and moving on to bigger animals like fish, frogs and mice.

'If I ever had any unease about it, I don't remember it,' she reflects.

At high school, her biology teachers stoked her fascination with science and she joined the biology club, enjoying trips to the mountains and sea to conduct field research. She was told that using animals in experiments was critical to life-saving medical advances and decided at a young age that this would be a good way to contribute to society. Her research career began straight after high school, when she participated in a study to test the effects of Vitamin E on rats who had surgically induced glaucoma. In 2006, she took up an opportunity to do a PhD in physiology at Case Western Reserve University in the US and was excited about her studies, which focused on cardiovascular disease and the relationship between a high-fat diet and heart failure. Then, midway through her research, which involved surgically inducing heart attacks in rats, Cheng's world turned upside down.

The laboratory where she worked had previously found that rats who had received the heart failure-inducing surgery and were fed a high-fat diet fared better than the ones fed regular rat food. This was surprising since it seemed to contradict what was observed in humans, and Cheng and her colleagues spent all their time trying to explain this phenomenon. They discovered some mechanisms specific to rats and submitted their findings to a scientific journal. However, one of the scientists reviewing their research suggested that their findings might only apply to rodents since it was known that rats naturally have high levels of high-density lipoprotein (HDL, sometimes called 'good cholesterol') and low levels of low-density lipoprotein (LDL, also known as 'bad cholesterol') regardless of their diet. Confused, Cheng sought answers from her thesis advisors but, to her great surprise, was advised not to pursue the issue.

'Your job as a PhD student is to graduate,' one of her advisors told her. 'It's not your job to think about this.'

Undeterred, she did some digging. She discovered that although LDL goes up after high-fat meals in humans, supposedly contributing to atherosclerosis, rats don't have a protein that increases LDL after high-fat meals, so her rat findings wouldn't have any relevance to humans. The research paper was slightly amended and published, but for Cheng, everything had now changed.

'I went from being proud and relieved that my first paper

got published,' she told me, 'to feeling shocked, confused and embarrassed that I had published a misleading study and harmed animals in useless research. I was also very angry that no one had told me about the downside of using animal models. This never once came up in the classroom, not even in the bioethics class. What happened to all those glorious claims that animal experiments save lives?'

Unsubstantiated claims

What indeed. Animal experiments are justified to the public on the grounds that they are essential for human health and medical progress, yet most animal research, in the UK at least, is not even intended to have direct application to human health. The lion's share (fifty-one per cent in 2021) falls into the category of basic research, which, as we saw in Chapter 2, aims to expand our knowledge of living organisms. A further proportion (twenty-one per cent in 2021) is conducted for regulatory reasons and involves testing products so they can be licensed for use. Only applied research (comprising twenty-seven per cent of all animal research in 2021) is intended to have direct relevance for health, but even this is not necessarily targeted at human health; in 2021, almost half of the research within the applied category was for animal diseases and disorders.[1]

Why, then, are we continually bombarded with claims about the value of animal research for humans? In response to increasing public opposition to the practice around the turn of this century, some prominent statements about its benefits began appearing. A particularly striking example was a statement first made by the UK's Royal Society – the oldest scientific academy in the world – which in 2002 read, 'Virtually every medical achievement of the last century has depended directly or indirectly on research with animals.' The statement was endorsed by many respected organisations and a version of it remains on the Royal Society website,[2] yet as Robert Matthews, Visiting Professor in Statistical Science at Aston University, has pointed out, it is inaccurate and entirely unsubstantiated; medical achievements may *involve* research on animals, but it does not follow that these achievements *depended* upon those animals' involvement.[3] Animals may have been incidental to any successes, or success might have been achieved *despite* the use of animals. As any statistician will explain, association is not the same as causation. Matthews further suggests that statements such as these should either be formally validated or replaced with statements capable of validation. This seems entirely reasonable, particularly if the statements are made by scientific organisations such as the Royal Society. In his important paper, Matthews also recommends the most appropriate statistical method for assessing the evidence generated by animal research, an approach that has since been widely adopted.

At around the same time, the Select Committee Inquiry on the use of Animals in Scientific Procedures, appointed by the House of Lords, concluded that, 'Animal experimentation is a valuable research method which has proved itself over time.'[4] Again, no evidence was supplied; the claim simply seemed to be that its longevity was sufficient proof of its efficacy. And in a controversial planning meeting for the building of a primate research centre at the University of Cambridge in 2003, Professor Sir Keith Peters, who was head of the School of Clinical Medicine at the time, argued that the need for the neuroscience research centre was 'self-evident'.[5] At the time of writing, Understanding Animal Research state on their website (again with no supporting evidence) that, 'Animal research has played a vital part in nearly every medical breakthrough over the last decade.'[6] When I contacted the organisation in 2014 to ask if they could supply me with evidence on the benefits of animal research, they sent me a link to four reports, the most recent of which had been published in 2006. But instead of rigorous, scientific evaluations of the benefits of animal research for human health, all four reports turned out to rely solely on expert opinion.

Expert opinion is not held in high esteem by those who understand evidence. Scientists often refer to a model known as the 'evidence pyramid', which places systematic reviews at the top and expert opinions at the bottom. Systematic reviews follow a strict protocol designed to minimise bias on the part of the reviewer; consequently, the evidence they generate is regarded as one of the highest forms available. By contrast, expert opinion is considered to be one of the weakest forms because it is subjective. Dr Aysha Akhtar, President of the Center for Contemporary Sciences, sums up the situation nicely in an online webinar:

'What you find is that the evidence in support of using animals in research to inform human health tends to come from the bottom of the pyramid, in other words, the evidence is in the form of expert opinions. On the other hand, the evidence in support of moving *away* from animal use, because it does not reliably inform human health, tends to come from the top of the pyramid, in the form of systematic reviews.'[7]

Isolated examples

It used to be that those making the case for animal research would recite a few isolated examples of where animal research had apparently contributed to human health[8,9] – the most popular being the development of penicillin, a vaccine for polio and insulin for diabetes – and those opposing animal research would carefully debunk these claims,[10] or epidemiologists and other scientists might suggest that the contributions were not as large as asserted.[11,12] Of course, animal research has been associated with medical advances, but the fact that these occur only occasionally suggests that they are random rather than the result of a paradigm that is able to consistently and reliably translate findings from animals to humans.

Pharmacologist Dr Bob Coleman began his career in the pharmaceutical industry in 1965, when he joined the Glaxo Group as a laboratory technician. His arrival coincided with a number of commercial successes for Glaxo, including the discovery and development of well-known drugs such as the bronchodilator salbutamol (marketed as Ventolin), ranitidine (Zantac), which inhibits the secretion of gastric acid (now discontinued) and beclomethasone (Becotide), among others. Virtually all of the research that resulted in the successful discovery and development of these drugs was conducted on laboratory animals; salbutamol, for example, relaxes airway smooth muscle in both guinea pigs and humans, and ranitidine inhibits gastric acid secretion in rats as effectively as it does in humans. Yet Coleman describes these successes as 'lucky', because other research programmes underway at the same time within Glaxo came to nothing. The reason for this, he writes, 'is simply that experimental animals have always been unreliable in their predictive power for human efficacy and safety, providing useful information on some drug candidates, but not on others'.[13]

If the animal research paradigm were undergirded by a robust underlying theory, animal experiments would be associated with medical advances consistently and reliably. In the absence of theory, however, we find ourselves on shaky ground. As philosophers Hugh LaFollette and Niall Shanks write, 'isolated examples explain or illuminate nothing. They become significant only in the context of well-confirmed scientific theories – theories such as evolutionary biology – which shape the way we conceptualize evidence.'[14] If the cases selected by animal researchers to demonstrate the value of their work were explained by a unifying theory, their arguments would be stronger. But no such explanatory theory exists. Indeed, the examples of failure selected by those opposing animal research are on a firmer footing because they can be unified and explained by evolutionary theory.

So how should we evaluate the use of animal models for human health? In a 2016 lecture to Understanding Animal Research, Sir Mark Walport, the UK government's Chief Scientific Advisor at the time, challenged the animal research community to submit their claims to scrutiny by conducting a 'Cochrane-standard review of the contribution of animal research to advances in health and well-being over the last twenty years'.[15] ('Cochrane' refers to an approach that relies on robust evidence such as systematic reviews.) None rose to his challenge, however, and scientists using animals continue to draw upon selected examples when making claims about the benefits of their research.[16] To date, no robust, comprehensive evaluation of the value of animal research for human health has been undertaken, but in some ways this is not surprising, as it would be a massive endeavour.

In its absence, then, we are obliged to piece together disparate bodies of research to gain an understanding of the contribution that animal studies make to human health. In this, the first of three chapters exploring this contribution, we begin by examining the relationship between animal research and the clinical sciences, those sciences involving human health and medicine.

From bench to bedside?

The traditional view is of animal research informing and feeding into the clinical sciences, neatly encapsulated in the phrase 'bench to bedside'. This conjures up a vision of scientists making discoveries at their laboratory benches and rushing these, in the form of beneficial treatments, to patients in their hospital beds. But the relationship between animal laboratories and human medicine is not as straightforward as it may seem.

Medical historian Dr Brandon Reines (also a veterinarian and Assistant Professor of Biomedical Informatics at the University of Pittsburgh) has explored cases from the fields of physiology, bacteriology, immunology and pharmacology. He observes that doctors frequently make medical discoveries in the course of working with, observing and researching their patients and that these discoveries are often 'dramatised' at a later date by scientists using animals in laboratory experiments. However, when historical accounts of these discoveries are written up, argues Reines, it is the animal experiments that gain the credit, rather than the preceding clinical research, insights and observations.[17]

Reines' theory is that animal experiments somehow dramatise existing clinical hypotheses rather than drive innovation. In support of this, he gives the example of Claude Bernard, the nineteenth-century French physiologist whom we met in the first chapter. Bernard claimed to have made many chance discoveries using animal experiments in his laboratory, giving the impression that this is where proper science took place, rather than clinics and hospitals. According to Reines, however, there is little evidence that this was the case. Although Bernard conceded that physiological investigations always start with observations from doctors and from pathologists conducting autopsies, Reines notes that when tracing the origins of his own discoveries, Bernard usually neglected to mention the clinical sources of his inspiration. In his memoir on the pancreas, for example, he claimed that the results of his laboratory experiments on rabbits led him to search for equivalent natural experiments in the clinical literature. However, the natural experiments conducted by doctors had been published many years prior to his rabbit experiments. As Reines puts it, Bernard had to go quite far out of his way to put the cart before the horse.

Animal experiments are more dramatic and memorable than clinical hypotheses and observations, proposes Reines, so when the histories of medical discoveries are written, it is the animal experiment that is reported as the defining moment, as with the 1949 'discovery' of the anti-manic action of lithium. This is generally reported as resulting from the psychiatrist John Cade's experiments with lithium on his pet guinea pigs, yet

Reines shows that lithium's effects on humans had already been established from clinical research published decades earlier, research that Cade himself reviews in his classic 1949 paper on the topic.[18] Nevertheless, it is still Cade's experiment on guinea pigs that is cited, not the prior clinical evidence.

Professor Paul Martin, a sociologist at the University of Sheffield, provides a more recent illustration of Reines' theory. Contrary to the common view, Martin and his colleagues found that innovation in the field of regenerative medicine (a specialty concerned with the replacement or regeneration of damaged human cells, tissues or organs) was often driven by clinicians working with patients rather than by scientists in their laboratories. Echoing Reines, he concluded that laboratory science was in many instances concerned with 'validating already existing clinical knowledge'.[19]

Supporting this observation, scrutiny of the publication dates of research papers indicates that animal and human studies often run simultaneously rather than consecutively. For example, Janneke Horn, Professor of Neuro-Intensive Care Medicine at the University of Amsterdam, was surprised to find that animal studies of nimodipine (a drug intended for stroke patients) were conducted at the same time as the human studies, rather than before.[20] And a similar picture is found elsewhere, with animal and human studies being conducted concurrently or, in some cases, the human studies preceding the animal studies.[21]

While a proportion of animal studies are conducted after human studies to satisfy regulatory requirements for drug safety, others may be conducted because they are inspired by the findings from human trials or by the hypotheses and observations that doctors generate as a result of their work with patients. Sometimes, clinicians observe that a drug intended for one purpose has an effect on something else, which can inspire scientists to conduct animal studies. Sildenafil, the active ingredient in Viagra, for example, was originally intended to treat cardiovascular problems; it was only when nurses noticed that men in the early human trials all turned onto their stomachs when the nurses came to check on them, that its potential for treating erectile dysfunction became obvious. Sildenafil was then tested in animals for this purpose.[22]

In other cases, a drug with a known effect on one condition may be explored to see if it can be repurposed for another. After the thrombolytic tPA, for example, was repurposed for use with stroke patients following its success with heart disease, scientists began to administer tPA to animals that had had a stroke artificially induced. Such studies may be conducted to 'validate' results found in humans (of which more later), or to investigate a drug's mechanisms or to explore different dosing options. The point is that in many cases, human research and clinical observations seem to drive and inspire animal research, rather than the other way round.

Translating from animals to humans

Some researchers have used systematic review methodology in an attempt to understand whether the findings from animal experiments translate to humans. One such study compared the findings from animal and human studies of six different drugs. The thinking was that if the findings from the animal and human studies agreed, then the animal studies were correctly translating to the human situation. The study found that the animal and human studies agreed for three of the drugs and disagreed for three.[23] But what does this tell us?

Unfortunately, not a great deal. First, animal studies can only inform human studies if they *precede* them, but this was not always the case; many of the animal and human studies ran concurrently. Second, where the findings from the animal and human studies agreed, it may have been because the animal studies were confirming findings already established in humans. Third, the quality of all the animal studies examined was judged to be poor, indicating that the animal data were untrustworthy. For all these reasons, it is hard to draw firm conclusions from this study. In 2019, scientist Dr Cathalijn Leenaars, from Hannover Medical School, tried a different approach. She and her colleagues conducted a comprehensive search for studies that investigated concordance between human trials and animal experiments. This time, the findings were more conclusive. After carefully scrutinising the studies, she and her colleagues concluded that translation from animals to humans appeared to be random; moreover, they were unable to identify any factors that increased the likelihood of translation.[24]

Material for a vast echo chamber

The wildly unpredictable nature of animal to human translation may go some way towards explaining why doctors and clinical researchers so rarely draw upon the findings generated from animal experiments.

In 2000, Dr Jonathan Grant from the RAND Corporation (a non-profit organisation that aims to improve policymaking) studied the evidence that was used to support fifteen different sets of UK guidelines aimed at helping doctors manage their patients' diseases. He and his colleagues analysed the publications that supported each guideline to see whether they reported laboratory or human research. Of the 1,761 papers they analysed, only four reported laboratory research, suggesting a minuscule, practically negligible, influence of animal and other basic science studies on clinical practice.[25]

Twenty years later, psychologist Dr Constança Carvalho, from the University of Lisbon, conducted an analysis of animal studies of major depressive disorder. She and her colleagues identified 178 rat studies of major depressive disorder and found that these studies were cited by other researchers, mainly other animal researchers, 8,712 times. Fewer than ten per cent of these citations were in clinical papers, indicating that doctors and medical researchers were not making much use of the rat findings and casting doubt on the contribution of rat research to the understanding and treatment of human depression.[26] In a related study, Carvalho found that in the majority of cases, human biology-based research methods received more citations in subsequent clinical research papers on major depressive disorder than did studies using non-human primates.[27]

Disturbingly, Andrew Knight, Professor of Animal Welfare and Ethics at the University of Winchester, found that of a random selection of ninety-five chimpanzee studies, half were never cited by other scientists and only fifteen per cent were subsequently cited in medical papers.[28] Similarly, a study in the field of orthopaedic surgery found that of 243 studies, involving over 9,000 animals, forty-two per cent were *never published* and of those that were, thirty-eight per cent were either never cited or were cited only once.[29] And finally, of 411 animal studies of surgical procedures that were cited 6,063 times, the vast majority (seventy-nine per cent) of citations were by other animal researchers. Only twenty-one per cent of the citations were from clinical researchers, and the average number of citations in human research papers was one,[30] 'a deplorably low number', as orthopaedic surgeon Seth Leopold commented.[31]

Contrary to the widespread claim that animal research informs and drives medical research, then, citation analyses indicate that the knowledge produced by animal studies is drawn upon only rarely by doctors and clinical researchers. Massive numbers of animals are killed for studies that are, for the most part, used only by other animal researchers, supplying material for a vast echo chamber.

In the 1970s, John Gluck was a psychology professor working in the laboratory of the infamous Harry Harlow, who was conducting a programme of research into maternal deprivation and social isolation using non-human primates. After several years conducting animal research, Gluck took some time out to work in a psychiatry department, an experience he describes as chastening. Although he had previously held a strong belief in the 'primary importance' of animal studies, his time spent working alongside doctors and patients had a decisive effect on him. In a memoir charting his journey from animal researcher to animal advocate, he writes about his period in the psychiatry department: 'during the entirety of my extensive fellowship I had never once heard a single reference to an animal study as providing any theoretical or treatment insight. This is not to suggest that my supervisors did not believe that animal research contributed to clinical understanding, but it was clear that they did not read or reference the animal literature in their day-to-day clinical teaching or practice, and I had no reason to believe they were atypical. When I would bring up the results of animal model studies during supervision sessions, they listened attentively but then moved on.'[32]

Gluck's experience with patients and their doctors was pivotal in making him reassess the justifications he had previously used to defend his research on animals and ultimately influenced his decision to move away from animal research. Dr Frances Cheng, whose story opened this chapter, also left the world of animal research. Straight after obtaining her PhD, she joined People for the Ethical Treatment of Animals (PETA) where her work now involves persuading companies, universities and regulatory bodies to replace animal research with human biology-based approaches. In a moving film tracing her journey, she questions the purpose of all the animal experiments being conducted.[33] So, what *has* been the outcome of all this research? After all, the proof of the pudding is in the eating.

6

DESPERATE PATIENTS

It's a cold afternoon in 1996 and I am walking to a crematorium in North London. Leaves gather beneath the plane trees lining the broad road by the cemetery. Eventually, a large brick building comes into view and, entering a dark chapel, I sit alone in one of the wooden pews. I'm at the funeral of a man who has died in his fifties following a stroke.

I got to know Jim and his partner Andy* because at that time I was spending many of my days on hospital wards and rehabilitation units, observing how stroke patients were being looked after. I would tuck myself into a corner and watch quietly as the daily routines of doctors' rounds, physiotherapy, meals, medication and family visits unfolded around me. Jim was in hospital for many months, moving from acute care to rehabilitation and back a few times, but Andy visited him faithfully every day. After Jim's death, Andy invited me to

---

* Not their real names

dinner, and we talked of the many years he now faced without his partner.

Jim and Andy were just one of many families dealing with stroke that I met throughout this period. At a previous post in London's East End, my work had involved visiting stroke patients in hospital and then in their homes to understand how the stroke was affecting them. Over those years, I met hundreds of people with stroke and listened to some of their stories in great detail.[1,2,3,4] Their suffering, and that of their families, was painfully obvious. Many were left with paralysis, weakness, incontinence and other disabilities that completely changed their lives, rendering them dependent and frequently worse off financially. Some lost their memories, or their ability to speak, others would spontaneously burst into bouts of tears or laughter. Depression was common. Apart from basic medical and nursing care, rehabilitation and perhaps aspirin to prevent a further stroke, modern medicine had little to offer. Some, like Jim, died. A few lucky ones made a spontaneous recovery. The others were discharged into nursing homes or simply sent back to their own homes to get on with it.

Curing diseases?

Around that time, there were great hopes that a thrombolytic (clot busting) drug called tissue plasminogen activator (tPA) would be able to help people with stroke. Following its success in treating heart attacks, tPA went straight into clinical trials with stroke patients[5] and was eventually approved for use. The problem, however, is that it can only be given to a selected ten to twelve per cent of stroke patients,[6] and of these, only half will benefit.[7]

Enormous resources have been poured into other potential stroke treatments over the decades, most notably a class of drugs known as 'neuroprotectives', which aim to reduce the damage caused by stroke. But while over a thousand of these drugs have been tested in animals, often with encouraging results, not one has turned out to benefit humans with stroke.[8] Shockingly, this research continues to be funded, calling to mind the definition of insanity usually attributed to Albert Einstein, namely, 'doing the same thing over and over again and expecting different results'. Meanwhile, stroke remains the second leading cause of death worldwide and one of the leading causes of disability. One advance has been the development of a surgical procedure to restore blood flow that can be given to about ten per cent of patients whose stroke is caused by a clot. Research into *drug* treatments for stroke has been spectacularly unsuccessful, however. The drug approval rate is estimated to be 0.1 per cent[9] and the single 'success', tPA, was repurposed from the cardiovascular field.

How is it that animal research – widely touted as being life-saving and crucial to medical progress – can have produced so few benefits for patients with stroke after all these decades? And is the case of stroke unusual?

Actually, no. The 'success' rate for Alzheimer's disease is also appalling, at 0.4 per cent.[10,11] Alongside dementia, Alzheimer's disease is the leading cause of death in the UK, but there is no effective treatment despite 'mouseheimer's disease' having been cured many times over.[12] At the time of writing, five drugs have been approved that can temporarily help with some of the cognitive and behavioural symptoms of Alzheimer's disease, but they don't affect the long-term prognosis and are not effective in all patients.[13]

The success rate for drugs aimed at treating cardiovascular disease, the leading cause of death worldwide, is a little better, but still very disappointing, at 6.6 per cent.[14] Overall, the proportion of drugs that makes it through from the first trials in humans to being approved and licensed is 9.6 per cent.[14] Drugs for rare diseases have higher success rates, bringing up the average, but unfortunately, the common chronic diseases that affect large patient populations do less well. There are many factors that play a part in the high failure rates, but animal studies, which precede human trials, are acknowledged to play a key role due to their poor ability to predict human safety and efficacy.[15,16] All these failed trials are costly, not only to the animals used and the trial participants who are exposed to ineffective or harmful therapies, but also to society as a whole. Research resources are wasted; effort that might be better spent elsewhere is squandered, while patients continue to suffer from painful and debilitating conditions for which there are no cures.

The situation with cancer – the 'emperor of all maladies'[17] – is dreadful. In a powerful lecture delivered at the Royal Society of Medicine in 2022, oncologist and scientist Professor Azra Raza noted that the mortality rates for cancer that year were the same as in 1930, even after controlling for the effect of different age distributions in the general population. Furthermore, she pointed out that the incline and subsequent decline in mortality rates completely parallel the rise and fall in smoking, as well as anti-obesity campaigns and some screening measures.[18] Although there have been successes, including the treatment of some leukaemias and lymphomas, these unfortunately account for only a few thousand cases among the millions of people diagnosed with cancer. And although people with cancer are surviving longer now, mainly due to earlier diagnosis, the treatments are largely the same as those that have been used for decades: surgery, radiation and chemotherapy.

Practising at the Herbert Irving Comprehensive Cancer Center at Columbia University, Azra Raza specialises in leukaemia, cancer of the blood. She has seen forty cancer patients a week for thirty-five years and is dismayed at the limited options she still has to offer those who come to her clinic.[19] When she began her career in 1977, she treated her acute myeloid leukaemia patients with a toxic combination of drugs known as '7 + 3'. Forty-five years later, and over fifty years since President Nixon declared his 'war on cancer', all she can offer them is the same toxic cocktail.

'Our cancer treatment has remained Palaeolithic,' she exclaims. 'It belongs in the Stone Age. Giving chemotherapy to humans is literally like taking a baseball bat to a dog to get rid of its fleas. We have to do better. We have to take the blinders off our eyes and recognise there is something we're not doing right.'[20]

The drug approval rate stands at 5.1 per cent,[14] despite the billions poured annually into cancer research. Those drugs that are approved may prolong survival for a couple of months at most, while some have no benefit in terms of survival, and a large proportion may actually harm patients.[19] In an awful twist of fate, Azra Raza's husband Harvey, also a cancer specialist, was diagnosed with chronic lymphocytic leukaemia. In her groundbreaking book, *The First Cell*, she describes a morning in May 1998, when Harvey passed her the *New York Times* over coffee to show her an article about some new cancer drugs.

'Within a year, if all goes well', the article read, 'the first cancer patients will be injected with two new drugs that can eradicate any type of cancer, with no obvious side effects and no drug resistance – in mice. Some cancer researchers say the drugs are the most exciting treatment that they have ever seen.'

However, along with ninety-five per cent of experimental cancer drugs, these of course went on to fail spectacularly in humans. No game-changing drugs became available to treat Harvey, and he died at the age of sixty-one, leaving behind his wife and their eight-year-old daughter.[19]

'The big problem we have,' Azra Raza tells me over Zoom, standing bolt upright before a vast wall of books, 'is that those cancers which we can diagnose early we can cure by removing, burning or poisoning them, but those that are discovered later are so complicated that none of the strategies we have developed work.'

'Why is that?'

'Because things have become so complicated by then,' she says, 'and it's impossible to replicate that sort of complexity in any model system, even living ones like animals. With the advanced cancers, everything we give is just palliative.'

The poor ability of animal study findings to translate successfully to humans is not just confined to the leading causes of death and disability such as cancer, heart disease, Alzheimer's disease and stroke. It covers the full spectrum of suffering, including traumatic brain injury,[21] osteoarthritis,[22] multiple sclerosis,[23] Crohn's disease,[24] motor neuron disease,[25] rheumatoid arthritis,[26] asthma,[27] HIV/AIDS,[28] major depressive disorder,[29] Parkinson's disease,[30] sepsis[31] and type 1 diabetes,[32] to name but a few.

Clearly, then, there is enormous unmet need that animal research has failed to address. Surely, if animal research were as essential as we are led to believe, we might expect to see some effective treatments for our most common diseases? After all, this is how animal research is justified to the public.

Improving health and longevity?

'Over the last 100 years biomedical research has contributed substantially to our understanding of biological processes and thus to an increase in life expectancy and improvement in the quality of life of humans and animals.' So reads the introduction to the Basel Declaration, a statement written by Animal Research Tomorrow, an international organisation representing the interests of scientists using animals.[33]

The claim that laboratory medicine, including animal research, is responsible for general improvements in health and longevity seen over the past century or so is common. So how well does it stack up?

Writing in the 1970s, epidemiologist and doctor Thomas McKeown did much to debunk this myth. He argued that the bulk of health improvement seen over the last three centuries was due, not to the contribution of medical interventions, but to a decline in infectious diseases which in turn was brought about by better nutritional status, sanitation and hygiene, control of water and food and behavioural changes that limited population growth.[34] Moreover, he maintained that mortality was already falling steadily due to public health measures, well before the introduction of medical interventions such as vaccines and antibiotics in the twentieth century. Where medical advances had occurred, he proposed, these were for relatively uncommon illnesses such as childhood leukaemia and certain cancers. For the vast majority of chronic conditions, there was no effective treatment, he observed, although he acknowledged the role of medicine in postponing death and offering symptomatic relief.

McKeown was criticised by clinicians, who suggested he lacked medical experience and was using old diagnostic categories in his analyses,[35] and by historians, who suggested he had misinterpreted the death records.[36] But few today disagree with his central conclusions, that curative medical measures played little role in the decline in mortality rates prior to the mid-twentieth century and that real, lasting improvements in the health of populations are primarily achieved by reducing poverty and improving social conditions. Indeed, philosophers Hugh LaFollette and Niall Shanks note that the 'upswing' in government support for biomedical research, and especially research using animals, did not begin until after the Second World War and that the largest increases occurred after 1970, by which time most of the improvements in lifespan had already occurred. Only a relatively small increase in lifespan occurred after the big increase in biomedical research, they observe, suggesting that while medicine may have *contributed* to the decline in mortality, 'it would be a gross mistake to claim that scientific medicine (especially that derived from animal research) was a major causal factor in this decline'.[37]

Breakthroughs as a result of animal research?

I first met Dr Jarrod Bailey, now Science Director at Animal Free Research UK, in 2004. Bailey's PhD was in viral genetics and initially he had been interested in the causes of premature birth in humans. Gradually, however, he began to question the role of animal experimentation and eventually left academia to investigate this. In 2015, he set out to determine what proportion of animal research 'breakthroughs' reported in the UK national press actually translates into approved interventions for humans. With his colleague Professor Michael Balls, Emeritus Professor of Cell Biology at the University of Nottingham, he searched a media database for all reports of animal-based biomedical research for the year 1995. To be eligible, the reports had to be of a specific intervention that made claims about benefits for humans.

Twenty-seven reported 'breakthroughs' met these criteria, some of which made astonishing claims. For example, a report about animal research on ageing stated, 'Researchers have discovered a natural hormone produced by the body that could delay the effects of ageing ... the hormone could help to defer such characteristic problems of old age as wrinkles, muscle fatigue, rheumatism, bone fragility, memory loss and some cancers ...' The report's authors stated that the results so far in animals had been 'spectacular'. Nevertheless, after detailed analysis, it turned out that the majority of the twenty-seven reported 'breakthroughs' failed to translate into approved interventions for humans twenty years later. Twenty were classified as outright failures, three as partial successes and three as inconclusive. Only one had clearly resulted in human benefit, giving a success rate of four per cent and providing clear evidence that media reports of animal research should not be taken to imply any future human relevance.[38]

Prior to Bailey's research, in 2003, Dr Despina Contopoulos-Ioannidis and Professor John Ioannidis from Stanford University, identified 101 papers published twenty to twenty-five years previously that had clearly outlined a potential future application for humans. Sixty-four of the papers reported animal studies. The wife and husband team then followed up each of the 101 studies to determine whether the promising finding had resulted in human trials and/or clinical use twenty years later. By 2003, the much-trumpeted human findings had not emerged; only five resulted in an intervention licensed for clinical use and of these, just one had clinical impact, giving a success rate of one per cent.[39]

A slightly higher success rate was discovered by Canadian scientists Daniel Hackam and Donald Redelmeier. In 2006, they searched leading journals for highly cited animal studies published between 1980 and 2000. Of the seventy-six they found, each of which reported positive findings, only eight (10.5 per cent) had resulted in approved interventions for humans by 2007.[40] Looking at the issue from a slightly different angle, scientist Dr Lindsay Marshall identified 916 animal studies of breast, lung and colorectal cancer that were funded by the US National Institutes of Health between 2014 and 2019. As a way of determining whether any of these 916 studies had an impact on human health, she and her colleagues investigated how many were associated with a human trial. Astonishingly, they found just twelve. That's 1.3 per cent.[41]

Research conducted with basic scientists – those conducting laboratory based, curiosity driven research, including animal studies – suggests that they tend to be motivated by the potential for scientific discovery rather than the possibility of applying their findings to human medicine,[42] and this certainly seems to be borne out in practice. In 2011, policy expert Dr Steven Wooding, now at the University of Cambridge, analysed the 'payback' from a random sample of twenty-nine research studies on cardiovascular disease.[43] Fourteen of the studies were categorised as clinical, i.e. human-based, and fifteen as basic. He and his team found that all twenty-nine studies had an academic impact, for example, they resulted in publications in scientific journals. But while each of the fourteen clinical studies also had an impact beyond academia, for example, they informed policy or resulted in health or economic benefits, only nine of the fifteen basic studies had this wider impact. Wooding's team went on to conduct a payback analysis of research into schizophrenia and found a similar pattern.[44] It concluded that, in terms of generating health, social and economic benefits, clinical research – in other words, research conducted with humans – has a substantially greater impact than basic research. The basic sciences produce academic knowledge, but this does not necessarily translate into wider benefits. On the basis of three different payback analyses, Wooding's team now gives this advice to funders: 'For greater impact on patient care within 10–20 years, fund clinical rather than basic research.'[45]

This would seem to make sense. Professor Peter Rothwell, a clinical neurologist at the University of Oxford, has convincingly argued that most of the major therapeutic advances in recent decades have been down to clinical innovation rather than laboratory-based research, and this despite clinical research being chronically underfunded.[46] The decline in stroke incidence and mortality, for instance, appears to be due to the development of stroke units,[47] prevention[48] and advances made as a result of managing patients in clinical trials[49,50] rather than to laboratory-based research. Clinical research has always been a crucial driver in producing new knowledge and translating it into routine practice, yet its contribution is frequently overlooked.[51,52]

It just works

Hugh LaFollette and Niall Shanks suggest that nobody, biomedical scientists included, knows how to evaluate the impact of animal research on humans. They report having asked animal researchers for a clear and precise defence of the practice, but none was forthcoming. 'If we were pressed to identify the most widely held, though not generally articulated, justification for the practice', they write, 'we would pick out some version of the "it just works" argument'.[37]

Obviously, 'it just works' is neither good enough nor true. And while there is no single, comprehensive evaluation of the impact on humans of animal research, there is plenty of evidence that can be brought together to address the issue, as we have seen. Like a jigsaw puzzle, when the disparate pieces of data are combined, they form a picture. And that picture does not look good. So little is received in return for the huge resources we invest in animal research. Certainly, we gain academic knowledge about biological processes in animals, but it has not been possible to reliably and consistently translate this knowledge into benefits for humans. There is a lack of robust, systematic evidence to support the frequent claim that animal research informs human medicine, yet there is plentiful evidence to the contrary. Despite the popular view, animal research is neither driving medical progress nor contributing to the development of treatments for our most common and debilitating diseases. Indeed, it is failing spectacularly in these respects. Does it fare any better at ensuring the safety of new medicines?

7

THE REAL GUINEA PIGS

In the spring of 2006, Rob Oldfield was thirty-one and in perfect health. Recently returned from an acting course in LA, a friend of his suggested that participating in a medical trial would be an easy way to make some money. An American company, Parexel, was offering volunteers £2,000 to test a drug intended to treat a form of leukaemia, as well as multiple sclerosis and rheumatoid arthritis, by modulating the immune system. Rob was attracted by the possibility of making a scientific contribution and had gained the impression that the risks were minimal.

'I was interested in the scientific contribution I could be making. It was a medicine being tested in a laboratory approved by the government. What could go wrong?'[1]

The experimental drug, TGN1412, had already undergone extensive tests in animals. In particular, it had been tested in cynomolgus and rhesus monkeys because of their relatively close relation to humans and had been found safe at doses 500 times higher than the dose given to the trial volunteers. This was the first time the drug would be tested in humans.

On the ward at Northwick Park Hospital, London, the atmosphere was light-hearted and relaxed. Eight healthy men, aged between nineteen and thirty-four, lay on their beds and received their doses ten minutes apart, watching as the medication slid down the clear tubes and into their veins. David Oakley, newly engaged and in the middle of planning his wedding, was first. Within minutes of receiving the drug, he had a major headache, followed by severe pain in his lower back. He twisted and turned, trying to find a position that was less painful. Then, almost immediately after receiving his dose, Rob realised something was terribly wrong.

'My whole body went freezing cold and I started shaking,' he recalls. 'This wasn't something you could stop, it was so extreme. It was horrendous.'[1]

Recent graduate Raste Khan, unaware that he was one of two men given a placebo, could only look on in horror at the scene unfolding around him.

'It was all manic, everything was happening all at once. They were vomiting, they were screaming in pain, people were fainting, they couldn't control their bowels ... it was like a horror movie.'[1]

Ryan Wilson, a trainee plumber who was taking part to fund some driving lessons, was rushed to the hospital's intensive care unit. The doctor in charge of the trial, Daniel Bradford, recalls the chaos.

'They tumbled like dominoes. One man tried to walk to the toilet and collapsed. The wards became chaotic, with blood, vomit, and staff and patients shouting. It was clear which two had been given the placebos.'[2]

All six men who took the drug were treated for multiple organ failure. All mercifully survived but were told they could suffer long-term disruption to their immune systems. Ryan Wilson lost all his toes and the tips of several fingers to gangrene.[3,4] A report into the disaster by the UK's Medicines and Healthcare products Regulatory Agency (MHRA) concluded that the serious adverse reactions experienced by the volunteers were the result of an 'unpredicted biological action' of TGN1412;[5] in other words, the animal studies had given no clue as to how the drug would behave in humans. Detailed follow-up studies by the National Institute for Biological Standards and Control found that there was a subtle but important difference in the way human and monkey blood cells process the drug, meaning that the monkey studies would never have been able to predict the catastrophic reactions suffered by the human volunteers.[6]

Only ten years later a similar story unfolded, this time in north-west France. Once more, six healthy men were taking part in a drug trial. Scientists hoped that BIA 10-2474 would treat a range of conditions including anxiety, chronic pain and neurodegenerative disorders such as Parkinson's disease. So far, the drug had been tolerated in humans at doses of up to 20mg; this trial was to assess its safety at a daily dose of 50mg.

After five days on this dose, one man became ill and was hospitalised with symptoms similar to a stroke. On the sixth day, four of the five other men were hospitalised with similar neurological symptoms, including headache, memory impairment and altered consciousness. Less than a week later, the first man to become ill lapsed into a coma and died. Two others were left with residual neurological impairment.[7,4] BIA 10-2474 had been tested in extensive animal studies – on mice, rats, dogs and monkeys – and at doses up to 650 times stronger than those given to the human volunteers. After the disaster, Dr Annelot van Esbroeck, then at Leiden University in the Netherlands, tested the drug on human cells. She and her colleagues discovered that the drug deactivated multiple proteins, causing disruption to the metabolism of human nerve cells, effects that had not been identified in the animal tests.[8]

Phases of drug discovery

Both the TGN1412 and BA10-2474 disasters took place during Phase 1 clinical trials. Also known as 'first-in-human' trials, these are conducted with a cautiously small number of healthy human volunteers. If all goes well in a Phase 1 trial, the experimental drug will progress to Phase 2 trials, which are conducted with small numbers of patients for whom the drug is intended. Phase 3 trials are conducted with larger patient populations. If a drug successfully passes through all three phases, the pharmaceutical company can apply to the medicines regulator for the drug to be approved for use in the wider population. In the post-approval stage, sometimes known as Phase 4, the safety of the new drug is monitored in the general population.

Prior to these human trials, however, a series of 'preclinical studies' are conducted. These may include tests on human cells and tissues, as well as computer-based studies or simulations, but to satisfy the requirements of regulators, they always include animal studies. Animals are used in this context to study drug toxicity, the argument being that whole living organisms are vital for this purpose, especially for new drugs about which nothing is known. Certainly, for a drug that has never been used before, it is useful to understand whether and how it interferes with a biological organism and at what concentrations. Yet, while we may gain a perfect understanding of how this new drug behaves in the whole living organism that is a rat, mouse or monkey, when we try to translate this information to humans, we come up against the familiar problem of species differences. The information may translate to humans ... or it may not. Species differences make it an unreliable and inherently risky process.

Animals are also used to identify which organs might be affected by toxicity and to work out a reasonably safe dose for testing the drug in humans. This is done by determining the dose at which no adverse effects are observed in animals and then converting this to a 'human equivalent dose' through a process of scaling according to body surface area. A safety factor is then applied, which involves dividing the human equivalent dose by a certain figure; the US regulator, the Food and Drug Administration (FDA), for example, uses a default safety factor of ten.[9] This safety factor is the only protection against any unexpected toxicities that may arise due to interspecies differences.

'What people don't realise,' Kathy Archibald tells me, 'is that we have no idea what a safe starting dose is. We just test it in animals and then reduce it by a certain factor, it's completely unscientific.'

Graduating with a degree in genetics, Archibald began her working life contributing to drug discovery and diagnostics within pharmaceutical and biotechnology companies. A great believer in modern medicine – indeed she reports that her life has been saved many times by antibiotics and surgery – she nevertheless became concerned after several of her friends and family suffered adverse reactions to medication. She started to look more closely at the way new drugs are tested, investigations that eventually led to the founding in 2005 of Safer Medicines Trust. The charity aims to reduce the number of future victims of inadequate drug safety testing by encouraging a transition from animal experiments to human biology-based tests in preclinical drug development.

'You just put the test substance into the animal, you know, it's like Caesar having his tasters to protect him from poisoning,' she says of toxicity testing. 'You can see the attraction, the deceptive simplicity. If the animal dies, we won't use the drug, and that will protect us, but animals are a black box – if it kills the animal, you think it's unsafe, but you don't know why. And of course, it might not be toxic for humans.'

Risks to patient volunteers

Disasters in Phase 1 trials tend to be newsworthy because they involve healthy volunteers, but things can also go wrong in the later phases of trials, when drugs are tested in patients. For example, a combined Phase 1/Phase 2 trial of the drug fialuridine, tested by the US National Institutes of Health in the 1990s as a potential treatment for hepatitis B, resulted in the deaths of *five* of the fifteen patients involved, with two others only being saved by emergency liver transplants. Toxicity tests in animals, including a six-month trial in dogs, had given the drug the go-ahead for testing in humans.[4,10] Because Phase 2 and 3 trials involve patients with existing health conditions, it is sometimes difficult to unravel whether adverse reactions are due to the experimental drug or underlying ill health. But unfortunately, since Phase 2 and 3 trials involve a greater numbers of volunteers, if anything does go wrong, it tends to be on a larger scale; Phase 2 trials can involve up to a hundred or so patients, while in Phase 3, the drug may be given to hundreds or thousands of patients.

The MHRA collects data on 'suspected unexpected serious adverse reactions' (SUSARs) that occur during human trials of medicinal drugs, but these data are not easily accessible. Following a freedom of information request from the *Daily Mirror* newspaper in 2014, however, the MHRA revealed that 7,187 clinical trial participants had suffered SUSARs during the period 2010 to 2014, over ten per cent of whom had died, although it could not be proven that their deaths were directly caused by the experimental drugs.[11] Adding to a lack of transparency in this area, researchers are not very good at reporting the adverse effects of drugs when publishing the results of their trials.[12] Nevertheless, it is clear that patients receiving experimental drugs frequently suffer serious, and sometimes fatal, adverse reactions, despite the preclinical safety tests. A brief look at drugs for two very common conditions, stroke and cardiovascular disease, illustrates this point.

In the field of stroke, the experimental drugs diaspirin, enlimomab, selfotel and tirilazad all increased the risk of adverse reactions and death for patients in Phase 3 clinical trials. Each of these drugs had improved outcomes in animals (e.g. reduced brain injury, improved neurological function), but each led to a greater number of serious adverse events and deaths in stroke patients who took the drugs when compared with those in control groups.[13,14,15,16] And between 1990 and 2012, sixty-three Phase 3 trials of drugs for cardiovascular disease had to be halted. Almost a quarter of these trials were stopped for safety reasons, including seven that were associated with an increased risk of death.[17]

One of these was torcetrapib, a highly anticipated drug that was intended to prevent heart disease and expected to be a major blockbuster. The trial recruited patients at high risk of heart disease and compared torcetrapib with atorvastatin, an existing medication for reducing cholesterol. By the time the study was terminated, there were thirty-four more deaths in the torcetrapib group compared with the atorvastatin group and a significant increase in major cardiovascular events such as heart attack and stroke.[18] Animal studies had suggested that the drug would have a beneficial impact on human cardiovascular health.[19] In this case, patients were not the only casualty; the fallout from the disaster contributed to the drug company (Pfizer) having to lay off ten per cent of its workforce.[20]

There are many such examples to be found across a wide range of different medical fields. In 2021, a trial for a much-touted gene therapy had to be stopped following the death of a child participant.[21] An earlier trial of the same drug, although at a higher dose, had to be halted in 2020 following the deaths of three children. The drug, AT132, was developed by Audentes Therapeutics for a rare paediatric disease caused by mutations in a single gene,[22] and trials began after it was found to improve muscle function and extend the lifespan of mice and dogs.[23]

Despite animal studies being conducted to safeguard humans, between seventeen per cent[24] and twenty-four per cent[25] of the drugs that fail in clinical trials do so due to safety issues.

Risks to patient populations

As noted earlier, once a drug has successfully completed the three clinical trial phases, a pharmaceutical company can apply to the regulator for their drug to be approved and released onto the market. In the post-approval stage, monitoring of the new drug continues as it begins to be prescribed to people in the wider population. Sadly, however, even licensed drugs can result in adverse reactions and deaths. Approximately half the drugs withdrawn from the market in Europe and the US are withdrawn due to safety issues.[26] Rofecoxib, for example, better known as Vioxx, was approved in the US in 1999 for the treatment of arthritis and other painful conditions. From 1999 until its withdrawal in 2004, there were an estimated 88,000–140,000 excess cases of serious coronary heart disease in the US,[27] many of which were fatal. One study found that people taking the drug were sixty-seven per cent more likely to suffer a heart attack in the two weeks after getting their first Vioxx prescription, compared with those who did not take it.[28] Worldwide, twice as many people were exposed to Vioxx as in the US, meaning the scale of the disaster was enormous. Yet Vioxx had a *protective* effect on the hearts of mice and other animals.

Troglitazone, approved in 1997 in the US for the treatment of diabetes, provides another example. The drug was withdrawn in 2000 after reports of deaths and severe liver failure that required transplantation. Animal studies had not detected its potential to cause adverse effects in humans, but an international study led by the US Evidence-based Toxicology Collaboration and the Norwegian Institute of Public Health found that tests on human cells and tissues showed strong indications of an effect on the liver, which could have clearly revealed the hazard.[29]

Adverse drug reactions (ADRs) – excluding those caused by prescribing errors – are estimated to kill more than 10,000 people in the UK[30] and 100,000 in the US each year.[31] Indeed a 1998 study calculated ADRs to be between the fourth and sixth leading cause of death in the US.[32] Studies conducted since then have not shown any decrease in the burden of ADRs in the US or elsewhere, while many smaller studies suggest the burden continues to grow. A large proportion of hospital admissions are also due to ADRs: 6.5 per cent in the UK and 3.6 per cent in the rest of Europe.[33] In the general population ADRs may be caused by interactions with other medicines as well as toxicities not predicted by animal studies, including rare adverse reactions that are difficult to detect until the drug is taken by a large number of people. Unfortunately, even if an ADR is rare, when millions of people are taking the drug, significant numbers will be affected. A study of ninety-three serious human ADRs concluded that only eighteen of them (nineteen per cent) could have been detected on the basis of the preceding animal data.[34]

Selected examples or consistent picture?

I have provided many examples of drugs that seemed effective and safe in animal studies but went on to cause serious and fatal adverse reactions in humans. But earlier, I argued that examples only become meaningful if they are set within a larger context, in which they illuminate a consistent truth or accepted theory. So, have I simply selected a number of failures from a sea of successes, or are my examples typical of a wider pattern?

In 2015, I met up with Dr Jarrod Bailey again, this time at The Kennel Club in Mayfair, where we were attending a debate about the replacement of animals in research. Frustrated at the absence of a robust, comprehensive evaluation of the use of animals in drug safety testing, he and his colleagues had just completed a series of analyses comparing animal and human toxicity data for over 2,000 drugs.[35,36,37] This was a first, because the sort of data they analysed is usually retained by pharmaceutical companies and rarely sees the light of day. Bailey told me what had motivated him and his colleagues to do the analyses.

'Really, just the number of times I was involved in conversations about the fact, and actually just my own frustration that there was no systematic published evidence to support or reject this. It hadn't been properly addressed. There was nothing to support this scientifically.'

A previous, much smaller study had found that the presence of toxicity in animal studies correlates with the presence of toxicity in humans,[38] but as Bailey points out, if a test is to be accurate and useful, it must be both sensitive (able to identify the presence of something) and specific (able to identify the absence of something). His analyses confirmed that the *presence* of toxicity in animal tests is indeed likely to correlate with the presence of toxicity in humans but that this correlation is neither reliable nor consistent. Importantly, however, he established that an *absence* of toxicity in animal tests was unable to predict an absence of toxicity in humans. In other words, if a drug appears safe in animals (i.e. no toxicity is detected), it can nevertheless go on to be toxic in humans. Or as Visiting Professor in Statistical Science at Aston University, Robert Matthews, put it to me, 'If Fido goes paws up, that's bad news, but if he's wagging his tail, it means nothing.'

Bailey and colleagues used a robust statistical method recommended by Matthews when analysing their data.[39] Although some of their findings were challenged by a 2017 study of 182 drugs,[40] they were subsequently confirmed in analyses conducted by pharmaceutical industry scientists on data sets containing over 3,000 drugs.[41,42] These large-scale analyses, then, verify that the examples given above, of humans being harmed by drugs found safe in animals, are not isolated or unusual but illustrative of the general inability of animal tests to safeguard humans.[43] And the theory that explains why animal tests are unable to reliably predict outcomes in humans is, as we know, evolutionary theory.

We are the guinea pigs

Human trials are of course necessary and will always involve an element of risk. However, the supposed intention of animal studies is to *minimise* that risk to humans, not to increase it. Yet, ironically, this is precisely the outcome if human trials are launched on the basis of misleading animal data.

An 'Investigator Brochure' is a document containing all the available evidence about a particular drug, which is presented to ethical review boards and regulators to help them decide whether or not a human trial should proceed. Unfortunately, the quality of evidence presented in these brochures is poor, indicating a failure to take the necessary steps to avoid bias[44] and the vast majority therefore reporting positive findings from their animal studies.[45] As we saw in Chapter 3, bias leads to untrustworthy findings and a tendency to overestimate a drug's effects. When he learned of the poor quality of the animal studies that led to clinical trials of treatments for motor neuron disease, Dr Francis Collins, then Director of the US National Institutes of Health, could hardly believe it. As he said to author Richard Harris, 'Humans were being put at risk based on that kind of data, and that took my breath away.'[46]

Patients are often desperate, particularly those with terminal illnesses, and they want to hang on to every last scrap of hope. Such people will often jump at the chance to enrol in a trial of a new drug. Yet Professor of Bioethics at the University of California, Mark Yarborough, notes that because poor-quality animal studies lead to over-optimistic assessments of experimental drugs, patients considering whether or not to participate in a clinical trial are likely to be misinformed about the potential benefits of new drugs.[47] They cannot be said therefore, to have participated in a proper process of informed consent, considered the bedrock of medical ethics, and are unlikely to realise that they risk making themselves even worse off than they already are.

'People think you're being protected and that going into

a clinical trial as a volunteer is not risky,' Kathy Archibald tells me, 'but this is not true. People are injured and killed in clinical trials, even in Phase 1. We need to accept that we *are* currently experimenting on people, that heroic clinical trial volunteers *are* volunteering to be guinea pigs, and without proper informed consent, because nobody tells them the risks they're taking. We need to acknowledge that.'

Lost opportunities

There is yet another way we humans may suffer as a result of animal tests, and that is due to the possibility of false positives. If a drug shows toxicity in animals, it is unlikely ever to be trialled in humans. This may seem like a good thing, but remember the issue of species differences; while the presence of toxicity in animals may correlate with the presence of toxicity in humans, it will do so inconsistently. And this means we might be missing out on drugs that could actually benefit us.

Many beneficial drugs might never have been used in humans had they been tested in animals first. Aspirin, for example, perhaps our most widely used drug, is toxic to rat embryos and rhesus monkeys[48] and to dogs and cats at some doses. Tamoxifen, an effective breast cancer drug, would very likely have been withdrawn from development had it been known that it caused liver tumours in rats, a fact only discovered *after* the drug had been on the market for years.[49] Similarly, gleevec, an effective treatment for a form of leukaemia, was almost abandoned because it caused liver damage in dogs. As we shall see in the next chapter, its development was pursued following success in human cells and early trials with leukaemia patients.[50] It is possible that some of our many untreatable human diseases lack therapies because potentially safe and effective drugs showed toxicity in animal tests.

The way forward

Although for many decades using animals was the best we could come up with, it is now time to extricate ourselves, suggested Gregory Petsko, Professor of Neuroscience at Weill Cornell Medicine, on America's National Public Radio:

'What I am saying is at some point you have to cut your losses. You have to say, "OK, this took us as far as it could take us, quite some time ago".'[51]

In the following section, then, we turn to a more hopeful scenario, to a science based on human biology and to a truly modern medicine that can harness groundbreaking technologies to develop and test drugs without using laboratory animals. This new science may even enable the use of drugs to be avoided in the first place.

PART THREE

BREAKING FREE

8

THE POTENTIAL OF A HUMAN CELL

Lorna Harries, whom we first met in Chapter 4, is warm, down to earth and gives the impression of being hugely capable. Professor of Molecular Genetics at the University of Exeter Medical School, and Chief Scientific Officer at biotechnology company SENISCA, she has dedicated her career to investigating the nuts and bolts of how human cells age. We chat in her university office, which is remarkably clear of the usual stacks of papers and clutter. Outside, young trees on the medical school campus are turning red and yellow in the early autumn.

Harries studies cells from human tissues and organs. She is one of an increasing number of scientists who have become disenchanted with traditional, animal-based methods and who have turned instead to more direct ways of investigating human disease and developing drugs. Discouraged by the low rates at which animal research translates into effective treatments for humans, these scientists believe that cutting out the 'middleman' – or animal in this case – will increase their chances of success.

'Only one in 5,000 new drugs actually makes it to being licensed for use,' Harries explains, 'and I just think that's a lot of money and time and a lot of animals and effort, and there must be a better way of doing it. If you are able to, it's best to research human diseases and human phenomena in human systems. It's immediately translatable.'

When a gene is activated, during development or in response to changes in the environment, it makes a message containing instructions to create the necessary proteins, the molecules essential for the proper functioning of the body. Harries' laboratory found that one of the things we lose as we age is the ability to regulate which messages get made. Her hope is that a better understanding of the mechanisms causing human genes to be activated or deactivated in disease and ageing will provide insight into how to treat the causes rather than the effects of disease. She explains that her current research evolved from a study in a human population.

'We asked the question: what changes as we get old? We thought we'd see lots of things. Not much, actually. Most things stay the same. But what does change is your cells' ability to decide what is made. You lose the molecular resilience of your cells to be able to say, "in this situation, I need to make this". So, we're trying to restore the ability of cells to make what they need to, so they can respond.'

Harries uses primary cells, which are as near as possible to the cells found in a living human being, and cultures them in human serum, the clear fluid in blood that contains many of the components necessary for cell growth. Human serum is more expensive than the bovine serum – harvested from unborn calves – that is typically used, but it makes the results more robust by avoiding the introduction of matter from another species. She shows me around her laboratory where scientists are working quietly and purposefully among a dizzying array of equipment and machines. Although I recognise the freezers, centrifuges and cell scopes, Harries has to explain what most of the technology does, showing me machines for studying cell energetics, sorting cells, analysing molecules and even one which she describes as a 'molecular photocopier'.

'It's a PCR machine,' she explains. 'It takes the very low signal from the genes we're looking at and multiplies it, so we can see what changes them.'

As we walk back to her office, she emphasises that it is the science that drives her use of new technologies.

'My motivation for moving most of my work animal-free is purely pragmatic,' she explains, 'purely driven by the cold, hard science, so I think it's quite nice that the ethics side fits in and everything, but I'm driven by the fact that a lot of the findings people are getting from animals just do not translate. We've cured cancer – I don't know what – thirty times in mice? If it's possible to do it animal-free, even if it's harder, even if it's more expensive, I'm going to do that because I'm happier that what comes out of it is going to be relevant.'

Human cells

Harries' work would not be possible without the decades of research into cell biology that preceded it, some of which Rebecca Skloot documents in her astonishing book, *The Immortal Life of Henrietta Lacks*. Skloot relates how doctors took a sample of cancerous cells from a young African American woman, Henrietta Lacks, just before her death in 1951 from cervical cancer. Lacks' cells, which came to be known as HeLa cells, were grown 'in vitro' (literally, in glass, i.e. a test tube) and were the first to survive in a stable state outside the human body for a significant period of time. But her cells not only survived; they thrived and multiplied, allowing scientists to learn what sort of nutrients and conditions are necessary to grow and manipulate cells in vitro. HeLa cells were responsible for an explosion of research into cell biology that not only built the foundations for current advances, but also provided vital insights into human health, disease and genetics.[1] The cells, which were taken without her consent, made substantial sums of money for some pharmaceutical companies, although notably not for Lacks' family which is now suing companies that continue to use the cells without compensating them.[2] Thankfully, the 2004 Human Tissue Act has since made it an offense to remove, store or use human tissue, or to analyse human DNA, without the donor's permission.

Cell lines such as HeLa are permanently established cell cultures that multiply indefinitely as long as they have the appropriate conditions. Because they can be maintained over days or months – indeed they are often described as immortal – they can be useful for studying the effects of drugs. However, there are quality issues with cell lines, since they may diverge genetically over time or lose specific functions when they are transferred from their original culture into fresh growth media for the purpose of propagating the line.

Primary cells, on the other hand, are taken straight from a living organ or system such as blood. While they don't last as long in culture as cell lines – usually only days or, rarely, weeks – their strength is that they closely represent the organ or system from which they are derived. So, in pharmaceutical research, the expectation is that drugs will work in the same way in primary cells as they do in the whole organ. Heart cells, then, stand in as a proxy for the heart, making them useful for understanding how drugs affect that organ. By testing drugs on primary human heart cells, scientists have been able to predict how drugs will affect actual human hearts and have shown that human heart cells are superior to canine heart cells in this respect.[3]

Likewise, liver cells are used to investigate how drugs affect the liver. Drug-induced liver injury, commonly known as DILI, is the most common cause of acute liver failure in the Western world. It's a major reason why clinical trials fail[4] and drugs are withdrawn from the market,[5] but DILI is difficult to predict and often only becomes apparent once a drug is taken by thousands of people. Primary hepatocytes, the cells taken directly from fresh liver tissue, offer the possibility of more accurately predicting liver toxicity and have become the mainstay of many safety tests routinely used in pharmaceutical laboratories. Research using human liver cells was able to retrospectively predict liver damage from fialuridine, the drug that killed five patients in a 1993 clinical trial,[6] and time and again, this type of retrospective research using human cells identifies information that would have been invaluable had it been done *prospectively*, before human trials. As we mentioned in the previous chapter, in vitro research conducted after BIA 10-2474 killed one healthy volunteer and left two others with neurological impairments found that the drug disrupted the metabolism of human nerve cells.[7] This effect had not been predicted by prior tests in mice, rats, dogs or monkeys, pointing to the vital importance of using human cell models to investigate the safety of experimental drugs.[8,9]

Human cells can be used to discover drugs as well as investigate their safety. In the 1990s, oncologist Dr Brian Druker was working on enzymes, the proteins that accelerate chemical reactions. He was particularly interested in BCR-ABL, the enzyme implicated in chronic myeloid leukaemia (CML), a type of blood cancer which at the time killed most people within five years of diagnosis. BCR-ABL causes the white blood cells to grow abnormally, crowding out the red blood cells that carry oxygen and overwhelming organs and suffocating tissues. Treatment options were either a risky bone marrow transplant or the drug interferon which caused nausea, pain and fever.

Druker learned about a compound called STI571 that could block the activity of enzymes. He got hold of a batch and added minute amounts to live white blood cells taken from a CML patient, as well as to some healthy cells. The cells taken from the CML patient died, while the healthy cells survived unharmed. Encouraged, Druker tested STI571 in mice, again with good results. However, when given intravenously to dogs, the drug caused blood clots, and when given in pill form, the dogs showed signs of liver damage. Druker was advised by some to drop the project. Instead, convinced by his work on the human cells, he went straight to the US regulator, the FDA, and in 1998, obtained permission to proceed with a human trial. After around six months of increasingly larger doses of STI571, the white blood counts of CML patients began to fall within the normal range.

'That's when we knew we had something the likes of which had never been seen before in cancer therapy,' recalls Druker.[10]

By 2001, his team established that STI571 caused the white blood cell levels to return to normal in the vast majority of CML patients. The drug – now known as gleevec (or glivec in the US) – was approved and has since become the mainstay of treatment for CML patients, saving thousands of lives each year. CML is unusual, however, because unlike most cancers which are caused by a multitude of complex interacting genetic and environmental factors, it is the result of a single aberrant protein, enabling scientists to focus all their efforts on that single target. Nonetheless, it demonstrates how knowledge of the biological functioning of a cell can lead to life-saving treatment.[11] If Druker had had less confidence in his human cell findings, or been swayed by the distracting dog studies, it might have been a very different story for CML patients.

Organoids

Cell lines have limitations because, essentially, they are two-dimensional, so more sophisticated 3D cell cultures known as 'organoids' have been developed. These are microscopic versions of parts of human organs that can self-assemble given the right conditions. They are formed of complex clusters of organ-specific cells and, because they often contain multiple cell types, they allow scientists to study inter-cell communication. Organoids are used for research into disease, as well as for drug development. Harries, for one, has developed an organoid formed of chondrocytes, the cells that line the joints, which her team is using to investigate osteoarthritis.

At the beginning of the Covid-19 pandemic, there were reports of infected people experiencing neurological symptoms such as delirium, memory loss and stroke, but it was unclear how or why this was happening. Professor Thomas Hartung, Director of the Center for Alternatives to Animal Testing at Johns Hopkins University in the US, had been using brain organoids to study how the brain was infected by viruses such as Zika. When Covid-19 struck, his team exposed the brain organoids, smaller than the head of a pin, to SARS-CoV-2, the virus responsible for Covid-19. Seventy-two hours later, the virus had infected the organoids and was multiplying inside them, illustrating the susceptibility of human brain cells to the virus,[12] a finding later confirmed by post-mortem studies and brain scans.[13] Organoids had shown for the first time how SARS-CoV-2 infects human brain cells,[12] something that animal studies had so far failed to do.[14]

Organoids commonly use human stem cells, those cells that are uniquely able to develop into specialised cells such as those of the blood, nerves, or liver in the developing embryo. Initially, stem cells were grown in the laboratory from early human embryos (often surplus to IVF procedures), but in 2006 a breakthrough made it possible to reprogramme some adult cells – such as those taken from blood samples or skin biopsies – to become more like stem cells. Known as 'induced pluripotent stem cells' (iPSCs),[15] these cells are also able to renew themselves and differentiate into different cell types. Organoids using iPSCs have proven invaluable in preclinical drug development, enabling scientists to study the progress of disease and the actions of drugs, including toxicity. A system using multiple organoids, for example, was used to test six drugs that had been recalled due to adverse effects in humans, and for almost all the compounds, it was able to demonstrate toxicity at human-relevant doses that animal studies had failed to detect.[16]

Amazingly, organoids also offer the possibility of personalised medicine. Samples of a tumour taken from a person with cancer, for instance, can be used to generate an organoid on which different chemotherapies can be tested to identify which is most likely to help that individual.[17] It is also possible to create organoids from the cells of people with rare diseases or conditions. In one case, a patient with cystic fibrosis was not responding to the usual treatment because of a rare gene mutation. Scientists generated an organoid from the person's nasal cells and then exposed it to a range of experimental drugs with the result that a therapy was identified that was able to successfully treat the patient.[18]

Organ-on-a-chip

London's Design Museum contains a tiny, remarkable piece of technology, so arresting that it beat six other runners up, including a self-driving car and a system for clearing the oceans of plastic waste, to become overall winner of the museum's 2015 Design of the Year award.

'They identified a serious problem: how do we predict how human cells will behave,' Deyan Sudjic, Director of the Museum, said of its inventors. 'And they solved it with elegance and economy of means, putting technology from apparently unrelated fields to work in new ways.'[19]

The winner was an 'organ-on-a-chip', designed by American cell biologist and bioengineer Professor Donald Ingber. The chip is made of a clear, flexible polymer about the size of a computer memory stick. It contains microscopic hollow channels, each less than a millimetre in diameter, which can be lined with living human cells taken from an organ, and through which blood, air and nutrients can be pumped. In short, the organ-on-a-chip recreates the unique microenvironment that cells are exposed to within the human body and is even able to mimic mechanical forces such as peristalsis, the wave-like muscle contractions that move food along the digestive tract. Cells in a dish may behave differently to how they behave in a living, breathing body, but an organ chip provides them with a more dynamic, body-like environment, enabling them to closely mimic cells in their natural state.

'If we want to make cells happy outside our bodies, we need to design, build and engineer a home away from home for the cells,' suggests Geraldine Hamilton, former President and Chief Scientific Officer of Emulate, the company that Ingber co-founded in 2013 to manufacture the chips.[20]

As well as providing a suitable environment for cells, the organ-on-a-chip acts as a mini-laboratory. Because each chip is crystal clear, researchers can observe what is happening at the cellular and molecular level and extract data for analysis. They can study basic biological processes, investigate disease and test the safety and efficacy of drugs in real time. For example, a disease such as asthma can be introduced into the chip, and immune cells or drugs can be added, enabling scientists to observe the cells' reactions.

Ingber's first chip was a lung chip, which his team began to develop in 2007. They focused on the alveoli, or air sacs, recreating the lung's interface between blood and air and simulating the mechanical movements of breathing by applying a vacuum to parts of the chip. When they exposed the lung cells in the chip to pathogenic bacteria, the cells engulfed the bacteria as normal lung cells would.[21] When they used the chip to mimic pulmonary oedema (a condition caused by excess fluid collecting in the alveoli) and introduced a new drug into the chip, the drug completely inhibited the pulmonary oedema.[22] That drug, GSK2193874, is currently being tested in human trials.

Ingber's research caught the attention of the US Defense Advanced Research Projects Agency, or DARPA. In 2010, DARPA put out a call to scientists, asking them to develop innovative methods for testing new medical compounds. DARPA had no interest in using animal models, clearly stating that these were of limited relevance to humans.[23] The agency boldly proposed a completely new approach: the simultaneous study of ten or more interlinked organs-on-chips to simulate the workings of the human body. The concept of the 'human-on-a-chip' was born.

Tens of millions of US dollars were invested over the next five years, leading to two main funding awards: one for the Wyss Institute at Harvard University, co-founded by Ingber, and the other for the Massachusetts Institute of Technology. The Wyss Institute was also awarded a joint grant from the FDA and National Institutes of Health to develop a heart–lung 'micromachine' for safety and efficacy testing.[24] This funding kick-started an explosion of research into the organ-on-a-chip technology, such that in 2016, Dr Francis Collins, then Director of the National Institutes of Health, predicted that by 2026 organ chips would 'mostly replace animal testing for drug toxicity … giving results that are more accurate, at lower cost and with higher throughput'.[25]

Ingber's work on organ chips grew out of a conviction that the traditional drug development model was broken because of the inability of animal models to accurately predict human responses.[26] The Wyss Institute has now developed several organ chips, including for the kidneys and skin, and Ingber claims that many are better at mimicking human physiology, disease states and human responses to drugs – and are therefore more relevant for studying these aspects – than animal models.[27] Interestingly, he also notes that the chips are generating insights, replicating human responses and enabling discoveries that are not possible in animals due to differences in biological mechanisms. In 2022, I was lucky enough to attend a presentation given by Ingber at the Animal Free Research UK conference in Birmingham, where he dazzled the audience with his output and received a Pioneer Award for his trailblazing work.

When Covid-19 struck, Emulate collaborated with the FDA and researchers all over the world. Human upper airway cell culture models were used to investigate how the SARS-CoV-2 virus infects the cells of the upper airways, and organs-on-chips were applied to evaluate the effects of new and existing drugs and to throw light on how infection affects the heart, kidneys, gut and brain.[28] Ingber's team employed their human airway chip to identify existing FDA-approved drugs that could potentially be repurposed for treating or preventing Covid-19.[29] The ability to do these sorts of investigations in weeks rather than months, as would be the case with animal studies, was vital in the context of the pandemic. Indeed, Covid-19 galvanised the team to quickly integrate its capabilities and bring its full force to bear on the challenge.[30]

Organ chips are beginning to be used more widely in drug discovery and development where they are valued for their ability to elucidate mechanisms of direct relevance to humans. The chemotherapy drug cisplatin, for example, is used to treat a number of different cancers, but it can only be taken for a limited amount of time due to its toxicity. A kidney-on-a-chip was able to replicate the toxicities induced by cisplatin through a human-specific mechanism not found in animals.[27,31] Further information was provided by scientists from Hebrew University who used organ-on-a-chip technology to discover that cisplatin prevents the kidney cells from releasing the glucose they absorb, causing a massive accumulation of sugar in the kidneys. The cells then start accumulating fat, which causes damage.[32]

Suspecting that this damage could be prevented by inhibiting glucose reabsorption, the scientists introduced cisplatin to the chip alongside the diabetes drug empagliflozin, which limits the absorption of sugar in the kidneys. The diabetes drug reduced the build-up of fat, so the scientists looked for evidence in humans to confirm their findings. They found patients who had diabetes as well as cancer and were therefore taking both cisplatin and empagliflozin. These patients had fewer signs of kidney damage, confirming the protective action of empagliflozin and demonstrating that the drug could be used in a new way to reduce the side effects of cisplatin, enabling cancer patients to tolerate treatment for longer. Professor Yaakov Nahmias, Director of Hebrew University's Grass Center for Bioengineering, was delighted, telling *The Times of Israel:*[33]

'Getting a drug to the point of clinical trials normally takes four to six years, hundreds of animals and costs millions of dollars. We've done it in eight months, without a single animal, and at a fraction of the cost.'

Data derived from organ chips have been used elsewhere to gain regulatory approval for a clinical trial. Biotech company Hesperos made a human tissue chip modelling the characteristics of chronic inflammatory demyelinating polyneuropathy, a rare autoimmune condition that causes muscle weakness. When French pharmaceutical group Sanofi applied their drug of interest to the chip, it restored neuronal function, enabling a clinical trial to proceed.[34] According to Dr James Hickman, Chief Scientist at Hesperos, of 7,000 rare diseases found in humans, only 400 have animal models, making rare diseases fertile ground for models based on human tissues.

'We're not just talking about replacing animals or reducing animals,' he says. 'These systems fill a void where animal models don't exist.'[34]

In recent years, organ chips have been used to retrospectively identify drug toxicities that animal studies failed to detect and to throw light on the reasons why some drug trials failed. A blood vessel chip, for example, was able to accurately predict thrombosis induced by monoclonal antibody drugs[35] and liver-on-chip technology revealed that rezulin, a drug for type 2 diabetes that had caused unexplained liver damage in clinical trials, caused liver stress even at low concentrations and before any damage was visible.[36] Then, in 2022, a landmark paper was published, reporting a study that used 870 liver-chips to test twenty-seven drugs that had been judged safe for human use based on animal study evidence but which had gone on to cause serious adverse reactions in humans, including liver failure and death. The liver chips were able to detect toxicity in almost seven out of every eight drugs that were toxic to the human liver, far outperforming tests in animals and organoids.[37] How much harm might be averted if organ chips were used more widely in drug development and testing?

One of many tools

Nevertheless, as with any methodology, in vitro research has its limitations and there are many ways in which it may be compromised. There might be batch-to-batch differences with the cells used in organ chips and organoids, which can create confusion when interpreting results. Furthermore, each laboratory may use its own particular recipe for culturing cells, making comparisons difficult, and there may also be variations in the cocktail of substances used to support cell growth outside the body, which can confuse the interpretation of findings. Even the use of plasticware can affect the results of experiments. Guidelines have been established to encourage good practice because all these issues need addressing if the research is to be valid and reliable.[38]

It is also the case that, because the field is new and developing, there are many different organ chips and approaches which can result in a lack of standardisation, making comparisons difficult. Moreover, organ chips cannot fully replicate the body's immune response, endocrine system or gut microbiome, nor many complex human responses and functions, such as those involving cognition. Nor is the human-on-a-chip a reality; challenges remain in terms of combining different cell types, the absence of a universal cell culture medium to connect different organs and the sheer multitude of factors that have to be taken into account when replicating a whole system response. It seems wise, therefore, not to overpromise.

'There are problems; there are always problems,' Lorna Harries tells me, 'but there are lots of little solutions that you can put together. There's not a one-stop shop for this.'

She is right of course. Human cell technologies on their own will not replace animal research. Just as different tools are used for different purposes and have different strengths and weaknesses depending on the job at hand, so too the various human biology-based approaches and technologies need to be used together to complement each other and compensate for what each may lack. In the following two chapters, we explore what else is to be found in this toolbox.

9

UNLEASHING THE POWER OF COMPUTERS

Denis Noble was a research student in his twenties when he was given permission to use the big mainframe computer in the basement at University College London (UCL). The year was 1960. Because he was only granted access to the computer between the unpopular hours of 2am and 4am, he conducted his experiments during the day and did the computing at night.

'When did I sleep? Well, for about three days, I probably didn't,' he recalls in a 2021 interview with Archives IT.[1]

With his supervisor, the physiologist Otto Hutter, Noble had made some discoveries about cardiac ion channels, the proteins in cell membranes that control the flow of electric current in the cell. Noble wanted to use these discoveries to develop the first mathematical model of heart cells. Initially, he planned to do this with a hand calculator but had to abandon the idea when he realised it would take him six months to do a single calculation. The Ferranti Mercury computer in the basement at UCL was very slow but still capable of doing each calculation in two hours rather than six months. Having given up mathematics at school, Noble admits feeling like a bit of an imposter when he sought permission to use the machine.

'It's hard to imagine nowadays,' he reveals in another interview, this time with philosopher Stephen Carney, 'when computers are everywhere, even in a six-year-old's bedroom, that in those days a computer was a god-like thing – there was only one, it was in a basement, it was lauded and revered, and the computer scientists went around with long beards looking like priests guarding their fantastic treasure.'[2]

Nevertheless, permission was granted and Noble taught himself the Mercury Autocode programme language. After several weeks spent debugging the programme, the results started coming in.

'They showed something fascinating, after each electrical pulse of this model heart which I was building, there was the hint that it would take off again; it was trying to do another pulse.'[1]

After some fine-tuning, Noble had a mathematical model that explained how the rhythm of the heart was generated. He published his findings in *Nature*, an impressive feat for a young research student, and left UCL shortly afterwards to take up a tenured post as fellow of physiology at Balliol College, University of Oxford. He remained there throughout his career and is now, in his mid-eighties, Professor Emeritus of Cardiovascular Physiology at the same college.

'It's like a dream when I think back on it nowadays,' he muses. '60-odd years later, I still find it astonishing that I managed to do what I did.'[1]

Noble's research eventually led to the development of ivabradine, a medication that produces a gentler cardiac rhythm for people whose heart rate is too high, particularly during exercise. The drug is now used worldwide. But of equal importance was Noble's pioneering use of computational methods for simulating biological organs and processes. Over subsequent decades, computational biology developed and flourished as the speed and power of computers grew.

Computer processors are made of silicon, and so the use of computers in biology came to be known as 'in silico', sitting alongside the more familiar 'in vivo' (in a living organism) and 'in vitro' (in glass). In silico approaches are now used widely within the biological sciences, to build models of human cells, tissues and organs and, within the field of toxicology, to investigate the safety of drugs. For safety testing, the known characteristics of a drug, or type of drug, plus information about the biological system into which it will be introduced, and knowledge obtained from preclinical or clinical trials about how that drug works, can all be fed into an in silico model to predict the behaviour of that drug in humans. In this way, drug toxicity, including drug-to-drug interactions, can be investigated, as can 'adverse outcome pathways', the chain of molecular and cellular events that lead to adverse drug reactions.

Predicting drug safety and efficacy using in silico modelling

In the US, software has been developed to predict whether new drugs will cause liver injury and to understand the mechanisms that contribute to drug-induced liver injury, or DILI. The aim is to prevent human adverse drug reactions, decrease the demand for animal tests, reduce costs and speed up drug development. The software, known as DILIsym®, predicted that the migraine drugs telcagepant and MK3207 would be toxic to the human liver, a prediction that led to their development being terminated even though animal studies had failed to raise any significant safety concerns. Had only animal studies been used, telcagepant and MK3207 may have gone on to harm humans. Additionally, DILIsym® predicted that a related drug, ubrogepant, would be relatively safe for the liver.[3] This was confirmed in human trials and ubrogepant was subsequently approved by the FDA without any precautionary labelling regarding liver safety.[4] This provides an excellent example of the ability of in silico modelling to predict both the presence and the *absence* of toxicity, the latter being a particular challenge for animal studies as we saw in an earlier chapter.

New drugs have potential to injure not just the liver but the heart too, and each year, thousands of animals are used to try to predict whether or not this will be the case. However, the accuracy of these tests is limited due to species differences,[5] meaning that many drugs are approved, only to be withdrawn at a later date once they have been used in large human populations and found to be cardiotoxic.[6] Clearly, then, there is an urgent need for more accurate methods of prediction. One common and potentially dangerous adverse drug reaction is an irregular heartbeat, or arrhythmia. Dr Elisa Passini from the University of Oxford decided to try to predict this without using animals. Building on work conducted by Denis Noble at the same university, she and her colleagues took the innovative step of constructing a 'human in silico drug trial'. They tested sixty-two drugs on computer models of human heart cells and found that the computer model was able to predict the risk of drug-induced heart arrhythmias with eighty-nine per cent accuracy, comparing favourably with the predictions from previously conducted animal studies that were only up to seventy-five per cent accurate.[5]

'Our positive results give confidence in our model,' Passini said, on winning a major prize for her research, 'and demonstrate how human in silico drug trials could play a major role in the replacement of animal experiments in the near future.'[7]

The value of these trials, believes Passini, lies in their focus on human biology, which makes the results immediately applicable to humans, but also in their ability to significantly scale up testing. Thousands of testing scenarios can be modelled in silico, including different diseases, patient characteristics and genetic mutations, so that the huge variation in human responses to drugs is properly reflected. This is clearly an enormous advantage and something that will never be possible using animals.[7,8]

In silico trials can be used to explore the efficacy as well as the safety of treatments. An international group of researchers across an impressive array of different disciplines is aiming to generate virtual populations of stroke patients based on anonymised data from recent clinical trials involving people with stroke.[9] Alongside this, the same scientists are developing software that models the stroke itself, as well as treatments for stroke and the effects of these treatments. These models will be combined with the virtual patient populations, allowing virtual patients to receive virtual treatments which can then be evaluated. The group hopes that the virtual trials will provide a deeper understanding of how strokes occur and explain why current treatments have had only limited success.[10]

Predicting drug safety using artificial intelligence

Professor Thomas Hartung is sitting outside his summer house on the shore of Lake Maggiore. Director of the Centre for Alternatives to Animal Testing at Johns Hopkins University, he talks to me over Zoom about using the power of computers to advance human medicine, all without using a single animal.

'The computational power has become incredible, it's almost insane,' he says.

In 1996, the supercomputer Deep Blue beat world chess champion Gary Kasparov, having used data from over 700,000 games to learn more than 200 million possible moves. Twenty years later, the computer programme AlphaZero was able to play chess as well as the best human player after only four hours of training. After nine hours, it was outperforming the best humans. Hartung has been applying the same principles of machine learning, also known as artificial intelligence or AI, to predict the toxicity of chemicals, including pharmaceutical drugs, and tells me about a study he published in 2018.[11]

'We showed that using machine learning, AI, big data, we were better at predicting the properties of chemicals than if you just run the animal experiments. And we showed this with 190,000 examples.'

He points out that there are an estimated 350,000

chemicals on the market, a volume that makes it impossible to assess using animal tests.

'A computer can run them in a day and tell you which ones look suspicious, which you should take a closer look at. With animals, it would cost billions.'

The methodology is awaiting formal validation but Hartung reports that Australia has already accepted it and that regulators worldwide are interested.

While in silico models use data that we provide, AI *learns* from the data we give it. AI makes use of the vast amounts of data that are now available, not just in the scientific literature, academic databases or on the internet, but also those data generated by medical imaging or by new biotechnologies such as wearable sensors that can detect and report biological and chemical processes. We now have access to more data than ever before, and machine learning algorithms and AI software tools use these 'big data', allowing us to gain as much as possible from the knowledge we already have and to build on that knowledge by making new discoveries.

Just as with in silico modelling, AI is being employed to predict which pharmaceutical drugs will work safely in humans. Organ-on-a-chip technologies can test the effect of drugs on organs, but it is challenging to do this on a large scale. Bearing this in mind, Israeli company Quris developed an automated platform to test new drugs on hundreds of organoids, a platform they call 'patients-on-a-chip'. Testing a drug on a single organoid gives only limited information, explains CEO Isaac Bentwich, because different people react differently to the same drug. For this reason, his company tests thousands of drugs known to be safe or unsafe on 'male' and 'female' organoids with different genomic make-ups. The data generated are used to feed into and continuously retrain the machine learning model.

For AI to learn whether or not a drug is safe, the scientists at Quris believe that machine learning models need to routinely run thousands, and eventually millions, of patient-on-a-chip experiments. They anticipate that massive experiments will eventually be possible and at a fraction of the cost of traditional animal and clinical trials. Quris works with biologists, chemists, nanoscientists, micromechanics and robotics experts and is currently testing its model using a drug for Fragile X Syndrome, the most common inherited cause of autism and intellectual disabilities worldwide.

'This will be a test case,' says Bentwich, 'to demonstrate how our system can bring a drug to market in five years with millions of dollars, not 20 years with billions.'[12]

Drug company Merck is now collaborating with Quris, using the 'patients-on-a-chip' to test its new compounds. Danny Bar-Zohar, Merck's global head of healthcare research and development, says the new technology will allow them to assess the impact of a compound once it has been metabolised in the liver and after it has crossed the blood–brain barrier and reached the brain. It will also enable them to test drug-to-drug interactions prior to human trials, something Bar-Zohar describes as the 'Holy Grail'.[13]

Repurposing and rescuing drugs with AI

In a pandemic, when there is a need to rapidly identify safe and effective treatments, an obvious potential solution is to use drugs that have already been approved for human use in other medical conditions (known as repurposing) or drugs that have failed in one context but may have potential in another (rescuing).[14] As the safety data are usually already available, animal tests can often be completely bypassed, saving time, money and animal lives. The costs of taking a repurposed or rescued drug to market are estimated to be around $40–80 million, compared with $1–2 billion for the development of an entirely new drug.[14]

Don Ingber at Harvard University's Wyss Institute observed that at the beginning of the Covid-19 pandemic, efforts to repurpose drugs were haphazard, generating results that were equivocal and potentially dangerous. His team identified the need to address the issue in a more systematic and human-relevant way. They used a machine learning algorithm to sift through data generated from tens of thousands of known drug compounds, to identify those with potential to fight the SARS-CoV-2 virus. Among those identified was the anti-malarial drug amodiaquine which, using an airway-on-a-chip, was shown to inhibit SARS-CoV-2 infection.[15] At the time of writing, amodiaquine is being tested in clinical trials across twenty sites in Africa.

Virtual humans

Some argue that animal research will always be necessary in preclinical drug development because there is a need to study how a drug behaves in an entire living system, particularly for new drugs about which nothing is known. A key objection to this, of course, is that knowledge gained in animals, despite having come from a whole living organism, cannot be reliably applied to humans. On the other hand, in vitro approaches alone are unlikely to be able to fully replicate the complex workings of the human body. So, what is the answer?

The biologists and software engineers at CytoReason, a company based in Israel, are attempting to build a model of the entire human body, cell by cell and tissue by tissue, to create digital simulations of the human immune system and of human diseases.[16] Each time new data are added, the model improves as a result of machine learning so that its accuracy increases over time, helping it to represent the human body more and more precisely.

'We are building a digital, computational simulator of the human body that is so accurate it can be used to predict responses to drugs,' explains David Harel, CytoReason's co-founder and CEO in an interview for *The Times of Israel*.

'Scientists can then take a specific medicine and test it out. This lets scientists see directly how new compounds affect the human immune system. Our disease models do in less than an hour what would take a mouse experiment 18 months. And not only is it faster, it's better, because the goal, after all, is to help humans, not mice.'[17]

CytoReason can also produce the necessary data at a fraction of the speed and cost of animal experiments. Unsurprisingly, their models are increasingly being used by major pharmaceutical companies, including Pfizer, Roche and GlaxoSmithKline, to help them discover and develop new drugs. Pfizer, for example, developed a drug that seemed to inhibit CCR6, a protein thought to be involved in a range of autoimmune conditions. The company was unsure, however, which conditions and patients the drug would work best in. Such questions are typically answered after years of animal studies and human trials, but rather than going down this route, Pfizer's scientists used CytoReason's models to narrow down the drug's possible uses and were able to identify that it would work best with ulcerative colitis. The drug is now in clinical development.[17]

As data from more sources are included, in silico and AI models will become increasingly powerful, enabling scientists to explore and integrate the complex networks and interactions of the human body. The CompBioMed project, for example, is coordinated by a team based at UCL and uses massively powerful supercomputers and a vast array of medical, biological and epidemiological data to build complex biological systems, or virtual physiological humans. This huge collaborative endeavour, involving academic and industry partners all over the world, uses the systems in a broad variety of biomedical settings, including drug discovery and testing, medical device manufacturing, research and personalised medicine.[18] One of its associated projects uses the world's first 'exascale' computer, which is capable of billions of operations per second. 'Frontier', as the supercomputer is known, will be used to simulate blood flow in the brain in the seconds immediately following a stroke, with different scenarios of blockages in the arteries being modelled to investigate how this affects pressure on the artery walls.[19]

Virtual humans provide an outstanding illustration of the use of 'systems biology' and of the role of computers in enabling this approach. For decades, biology has adopted a reductionist perspective, breaking down living systems into their smallest components, the individual genes and molecules, and looking at them in more and more detail in order to understand them. But the challenge now is to integrate the information gained, or as Denis Noble says, we have to put Humpty Dumpty back together again.[2] Systems biology attempts to integrate what has been learned through the reductionist enterprise into a larger picture, understanding biological processes as interacting systems rather than isolated parts.[20] Noble believes that while science needs reductionists, it also needs those who can recognise large-scale patterns and see the whole picture.[21] Nevertheless, when it comes to computers reproducing the complexity of human cells, he is cautious.[1] So is Professor Azra Raza, whom we met in a previous chapter.

'Biology is based on a cell,' she tells me, 'and eighty to ninety per cent of the cell is liquid. Within that liquid are swimming one trillion molecules, within one cell, and they are constantly communicating with each other. They know exactly which proteins must be made, which must be shut down, which genes have to be turned on and off. Can you imagine the complexity of this intracellular signalling?'

Because of this complexity, Raza believes it unlikely that AI will ever be able to accurately replicate cellular intelligence. And there are other issues with 'virtual humans'. How accessible would they be, for instance? It seems unlikely that the NHS will provide everyone with a personal avatar for determining the best treatment options, so is it something that would only be available to the rich?

Nevertheless, even if the jury is out regarding how far we can go with virtual humans, in silico and AI models are already enabling large numbers of drugs to be tested quickly and cheaply – far more quickly and cheaply than is possible using animals — and of course more ethically. Furthermore, and importantly, these models appear to have greater accuracy than animal studies. As a result of using in silico modelling and AI approaches, the pharmaceutical giant AstraZeneca has managed to dramatically reduce its failure rate in the first stage of human trials. In 2011, thirty per cent of its drugs failed on safety grounds in Phase 1 trials, but in the seven years up to 2022, not one has failed.[13]

Perhaps the most promising approach is to use the best of both worlds, combining in silico and AI with human biology-based technologies, as Quris has done, opening the door to the creation of *real* intact systems as well as virtual ones.[22] Whatever happens, the scene is set for computational approaches to play a significant role in replacing animals in drug development, particularly drug safety testing. But what about avoiding or minimising our use of medicines in the first place? After all, prevention is better than cure.

10

SOMETHING NEW, SOMETHING OLD

Many years ago, I conducted some research with colleagues at the University of Bristol into the reasons why people don't take their medicines as prescribed.[1] The academic literature was full of bewilderment about why patients might ignore their doctors' prescriptions, a phenomenon that used to be referred to as 'non-compliance'. Combing through the literature, we discovered a widespread wariness and distrust of pharmaceutical drugs, with many people testing their medicines first to see whether they worked, had side effects or could be taken at a lower or less frequent dose. In short, we found strong evidence of a reluctance to take medicines and a preference to minimise intake.

This is hardly surprising, given widespread awareness of the potential dangers of medicines. As noted earlier, it is estimated that more than 10,000 people in the UK and over 100,000 in the US are killed by adverse drug reactions each year.[2] In previous chapters, we discussed how misleading animal studies can lead to ineffective or harmful medicines entering clinical trials or being approved for human use. But additional issues may affect the safety of medicines, including poorly designed clinical trials, medication errors, ineffective monitoring and the fact that medicines are usually prescribed using a 'one size fits all' approach.

Thanks to our increasing knowledge about the complete set of genes that makes us human – the human genome – it is now becoming possible to address the latter issue using an approach known as 'pharmacogenomics'. Taking a more personalised approach, scientists can discover an individual's gene variants, enabling them to work out which specific medication (and dose) is likely to suit that individual best.[3] The Human Genome Project, launched in 1990 and completed in 2003 by an international group of scientists, made this and many other scientific feats possible, as we shall see. The scientists studied the genome of a small number of diverse individuals and assembled these into a map representing the 'general' human genome, providing a detailed blueprint for building every human cell.[4]

The insights arising from the Human Genome Project and other scientific advances will not only make treatments safer and more bespoke but may even enable us to avoid taking medicines in the first place. Many believe that the coming decade will see a shift from reactive to proactive healthcare, in other words, a move from treating diseases to preventing them. If twentieth-century medicine can be characterised as focusing on the treatment of disease, there are certainly signs that twenty-first-century medicine is going to pay more attention to preserving wellness.

Preventing the silent death of brain cells

In the case of Alzheimer's Disease, chemical changes in the brain happen many years before the person or those around them begin to notice their effects. This raises the possibility of detecting these changes at their very earliest stages, perhaps through a blood test, allowing preventative action to be taken long before the disease takes hold.

Unfortunately, by the time we realised what was happening in my father's brain, the nerve cells involved in thinking, learning and memory were already being silently destroyed. Then, little by little, as the disease progressed, the nerve cells elsewhere in his brain began to succumb. We'd noticed a change soon after he retired. Always impeccably correct about his grammar – he'd been a journalist, BBC editor and literary agent – he suddenly began to lose and jumble his words. After he died, we found notebooks with words written over and over again in pencilled capitals, each with a slightly different spelling. On a scrap of paper, he'd practised my mother's name and a birthday greeting. He must have been aware of the letters slipping away, at least at first.

The police had to bring him home once or twice after neighbours found him wandering along the road out of the village. Then he began to return from the newsagent's, not with his usual broadsheet but with whatever tabloid appealed to him that day. As things got worse, he began to obsess about certain objects – his pipe, tobacco, Extra Strong Mints – and would accuse my mother of hiding them. He still recognised us, his children, but couldn't understand how we'd suddenly become so tall. In the morning, he put his pants on over his trousers and at night my mother would settle him into bed, only for him to get dressed again three or four times before dawn. On one of these occasions he mistook the stairs for the bathroom and fell all the way to the bottom, dying many weeks later of a subdural haematoma.

Alzheimer's disease and dementia have been the leading causes of death in the UK since 2011.[5] Alzheimer's disease is complex and caused by a combination of genetic and environmental factors, which make it all but impossible to mimic using animal models.[6,7,8] While a few drugs can temporarily improve some of the cognitive symptoms, none halts or slows the destruction of the nerve cells. Increasingly, it is argued that chronic, complicated diseases such as Alzheimer's are not caused by single factors that can be targeted by drugs but are the result of a *system* imbalance and so require a different approach.[9]

When his wife was diagnosed with the disease in 2005, American biologist Leroy Hood didn't take long to conclude that searching for a single cure was futile. Instead of waiting for Alzheimer's to show itself, he decided efforts should be directed at detecting the disease much earlier. Currently, Hood and his team of biologists, computer scientists, engineers and mathematicians at the Institute for Systems Biology are monitoring individuals at high risk for Alzheimer's, by means of regular genetic and blood tests, investigation of their microbiome (the community of microorganisms living in their gut) and health measurements generated from wearable Fitbit-type technologies. The hope is that the study will allow researchers to detect the very earliest signs – the biomarkers – of the transition from wellness to Alzheimer's disease, creating the possibility of intervening early on.[10]

This would be especially useful for those at a higher risk of getting the disease. A defective form of the gene APOE is associated with Alzheimer's. One copy of the defective gene – APOE4 – gives an individual a greater chance of developing the disease by the time they are sixty-five, but two copies of the defective gene make the likelihood much higher. Wanting to know if the rest of his family were likely to get the disease, he and his children took genetic tests. Hood found out that he – like his wife – had one copy of APOE4. Happily, his daughter had no copies of the defective gene but his son, unfortunately, had two.[10] Knowing their increased risk of Alzheimer's, what steps could Hood and his son now take to prevent the disease?

Because genetics alone do not determine outcomes but interact with environmental factors such as diet and lifestyle to create susceptibility to disease, Hood advocates intervening early to change the course of events. For those at greater risk, early 'personalised' interventions addressing the specific vulnerabilities of each individual will be crucial for maintaining a healthy brain, he suggests. A person following this approach would have their unique characteristics assessed through analysis of their blood and factors such as exposure to environmental toxins, as well as their diet, exercise and sleep patterns. Then a tailor-made treatment plan would be developed, consisting of supplements, exercise, dietary adjustments, drugs, removal of toxins and lifestyle changes.[11] There is some evidence that multimodal programmes like this can maintain, and even improve, cognitive functioning.[12,13] Hood's Institute is in the process of testing various combinations of therapies for patients with early Alzheimer's disease in clinical trials. If treatment turns out to be effective, he suggests families with a history of Alzheimer's might feel it worthwhile taking a genetic test to assess their risk.

Health preservation

In another project, 108 'pioneers', including Hood, had their genome sequenced and then, every three months over a nine-month period, had their blood taken to analyse 1,200 different chemical substances. In addition, their gut microbiome was analysed and the bacterial species quantified, while digital devices measured aspects such as activity, sleep, pulse rate and brain health. Ninety-one per cent of the volunteers were found to have nutritional issues, mostly due to genetics, while sixty-eight per cent had inflammatory difficulties which were mostly due to diet. Almost half were 'prediabetic'. Yet Hood writes that they 'were able to move the needle on nearly all of these measures in just nine months, making virtually all of the pioneers healthier …'.[14]

The project involved gathering detailed information about the volunteers using an approach known as 'deep phenotyping'. A phenotype is an individual's observable traits, but deep phenotyping goes beyond observable traits or what is typically recorded in medical records (e.g. weight, blood pressure) to collect the fine details about how a disease manifests at the deeper levels of organs, tissues, cells and molecules. This information is then integrated with other kinds of data, for example on the patient's genome, allowing researchers to look for patterns.[15]

Projects that generate such an enormous amount of information about each individual create 'data clouds' which, suggests Hood, will generate new ideas and theories about human biology and human disease[16] and eventually allow us to diagnose and reverse common chronic diseases well before they manifest.[14] An example of this is the All of Us research programme in the US, which is combining genomic data from thousands of individuals with data from their electronic health records, wearable software and survey responses, as well as data from the US Census Bureau about the communities in which they live.[17] The aim is to build 'one of the most diverse health databases in history', generating data on a wide range of different diseases, risk factors and responses to treatment and helping to move away from the 'one size fits all' medical approach.

The personal information Hood gained as a result of the 108 Pioneers project allowed him to correct five of his own nutritional deficiencies with supplements. One of these deficiencies was Vitamin D, which he tried to correct by taking a thousand units per day. When this achieved nothing, he investigated further, discovering that he had two gene variants that blocked the uptake of Vitamin D, meaning that a much higher level was needed to achieve and maintain a normal level.[14] Since Vitamin D deficiency is thought to be a risk factor for Alzheimer's disease, this information was of vital importance for Hood.

Viewing disease as the outcome of a complex interplay of genetics, environment and lifestyle, and intervening accordingly, has come to be known as 'functional medicine'. Within this paradigm, the focus is on understanding the individual's own particular blend of genes, environment and lifestyle, and on identifying and addressing the root cause as it manifests for that person.[18] It recognises that one condition may be the result of many system imbalances, and also, that one system imbalance may result in many different conditions.

Digital twins for preserving health

Many technologies are now becoming available to support or enable personalised approaches to healthcare, at least in theory. The Whole Body Digital Twin™ is a system built from thousands of data points gathered from wearable sensors and is described as a dynamic, digital representation of an individual's metabolism. Recognising that everyone's metabolism is different, the Whole Body Digital Twin app aims to deliver individualised guidance to help reverse and prevent multiple chronic diseases.[19] In October 2021, the company, Twin Health, secured $140 million to scale up its technology.

Another kind of 'digital twin' exists. This one is generated, not from data provided by wearable sensors but from known medical data about an individual, such as blood pressure, heart rate, weight and cholesterol level, as well as demographic data such as age, sex and race. Professor Gunnar Cedersund, from Linköping, Sweden, spent twenty years building in silico models of almost all the organs of the human body, as well as adipose and muscle tissue and blood circulation. On the basis of this work, he developed a personalised computer model that allows different events to be simulated and tested to see how an individual would be affected if, for example, they changed their diet or started exercising more.[20] The technology enables risk factors to be calculated and treatments tailored for the individual. Alternatively, a physician could use this digital twin to test interventions, for example, to investigate how an individual might respond to different drugs.[21]

On the whole, these digital twins are different from the 'virtual humans' used in drug development, because they are personalised. Having said that, the virtual humans developed by the huge CompBioMed project, discussed in the previous chapter, can also explore how changes in lifestyle might affect health, ageing and quality of life. Describing their virtual human, CompBioMed write: 'Imagine a virtual human, not made of flesh and bone, one made of bits and bytes and not just any human, but a virtual version of you, accurate at every scale from the way your heart beats down to the letters of your DNA code.'[22] The collaboration claims that the avatar will be used to simulate how drugs interact with an individual's unique genetic make-up, enabling a drug to be selected that will precisely suit that individual. It can even be used to do such things as calculate the mechanical forces on an individual's bones to predict the risk of fracture.[23]

Improving the prevention and treatment of cancer

A movement away from the 'one size fits all' approach is also evident in the field of cancer. Here, knowing more about a tumour's characteristics can give doctors a better idea of how that tumour will behave, enabling them to identify therapies likely to benefit their patients, as well as those liable to do them harm. Precision medicine, as this approach is often called, also helps health professionals match patients to clinical trials. In the UK, for instance, there are thousands of NHS cancer patients who are reasonably well but who have run out of treatment options. Now they are being offered the chance to have the specific genetic code of their tumour analysed from a blood sample, enabling doctors to select an experimental treatment that has a greater chance of helping them. A pilot scheme, run by The Christie NHS Foundation Trust in Manchester, found that patients involved in this initiative were more likely to see their tumours shrink. The scheme is now being rolled out across England, Wales, Scotland and Northern Ireland, with thousands of patients being recruited via eighteen cancer centres.[24]

These sorts of initiatives are likely to make real differences and they have nothing to do with animal research. They have developed from human-focused research and specifically from research into the human genome and the genetic make-up of human cancers. The UK's 100,000 Genomes Project, for example, sequenced the genomes of NHS patients with cancer and rare diseases and is already leading to the diagnosis and treatment of rare cancers.[25,26] The project aims to make genomic medicine a routine part of healthcare and is working with the NHS Genomic Medicine Service to deliver this.

However, most people die from cancer because the diagnosis comes too late, when the cancer is already too far advanced. This means early detection is vital if cancer survival rates are to improve. As I write, the UK's first Early Cancer Institute has just been launched. Aiming to detect the disease at its earliest stages, the scientists at the Cambridge centre will focus on cancers that are hard to treat, such as those of the lung and pancreas. But while the current approach is to focus on specific cancers, advances in screening mean that in the near future it may be possible to screen individuals for a much wider variety of cancers.

The Oncology Think Tank aims to shift our focus from treating a disease that has already got its claws into us, to preventing it from developing in the first place. Members propose focusing on the very earliest point at which precancerous changes can be spotted, using information contained in the blood, or obtained from wearable sensors, so that the danger signs of the disease can be detected far ahead of its bodily appearance.[27] One of the Think Tank's founder members is cancer specialist Professor Azra Raza, whom we met earlier. Almost four decades ago, she established a tissue repository and filled it with blood and bone marrow samples from thousands of her patients who were suffering from acute myeloid leukaemia and a group of preleukaemic disorders known as myelodysplastic syndromes. She collected samples from the same individuals at various time points, allowing her to monitor disease progression, pinpoint risk factors that create susceptibility to cancer and detect cancers at a stage at which they are easy to treat.[28] Now she is in a position to use this information to benefit people, explaining that a single drop of human blood can allow the signals, or 'footprints', of cancer to be detected many years before the disease becomes apparent.[29]

'We will be able to find the molecular markers, the protein metabolic markers, way before the disease declares itself,' she tells me over Zoom. 'I am completely confident and optimistic that it's going to happen.'

American healthcare company GRAIL shares her confidence. Using the samples in her repository, the company has developed a simple blood test that may be able to detect more than fifty different cancers at their earliest stages. The company is working with the NHS to pilot what it calls the 'Galleri test' in a trial with 140,000 patients.[30] The Galleri test is still in the early stages of development and large, long-term, randomised controlled trials will be necessary to demonstrate real-world benefits as well as lack of harm from false positives, which can potentially lead to invasive investigations and overtreatment. However, given its ability to identify cancers that are difficult to diagnose early, such as ovarian cancer, this test – if successful – could be a complete game-changer.

'This is how we are going to shift the paradigm,' Azra Raza tells me. 'Within the next ten years, we will go from active treatment to preventive treatment. We are going to anticipate. We want to monitor wellness to find illness before it has become clinically detectable. And then try to prevent it at that stage.'

Azra Raza shows me the Fitbit around her wrist, explaining that it is recording her blood pressure, heart rate, oxygen levels and all sorts of other dynamic information about her vital signs. Passionate about the potential of technology to monitor human health and wellness, she tells me that we can even be monitored at night by sheets that scan our bodies as we sleep, looking for 'hot spots', the areas where blood vessels grow to provide a blood supply for new tumour cells. She also points out that the treatment of a cancer that is detected early will be far less draconian than the treatment of an advanced cancer, which usually involves radiation, chemotherapy or surgery. Ninety to ninety-five per cent of research funding goes into studying disease that is already far advanced, she tells me, and this is mostly done using animal models.

'To study human diseases, we should study humans!,' she exclaims in exasperation.

Pancreatic cancer has non-specific symptoms and, like ovarian cancer, is notoriously difficult to identify until its later stages. This means that at present, half of those with pancreatic cancer die within three months of diagnosis. But if the cancer is found early, before spreading, the chances of successful surgery to remove it are greater. Researchers have reported that certain changes in the human microbiome – detectable in stool samples – are associated with pancreatic cancer in both its early and late stages, creating the possibility of detecting this cancer much earlier.[31] This is the value of human-focused research.

Tried and tested approaches

In 2013, neurologist Professor Peter Rothwell from the University of Oxford gave a lecture on stroke and dementia at the Oxford Biomedical Research Centre.

'I want to concentrate on what we can achieve just by getting the simple things right,' he began. 'So often, medical research is portrayed in the media as very technical – genetics and white coats and microscopes – and that's clearly important, and much of the funding ... is for that sort of basic science, but I think that sometimes we lose track of how much can be achieved just by doing the simple things right.'[32]

Rothwell was making the point that, frequently, we can achieve a great deal simply by correctly applying the knowledge we already possess. Although dazzling technologies and innovative approaches now exist to advance medicine, it is worth recalling that an impressive toolkit already exists for learning about humans and managing diseases.

To this end, Rothwell provided data from clinical trials showing that people with stroke cared for on stroke units as opposed to general medical wards are less likely to die, and less likely to be disabled or institutionalised, not because of some wonder drug or technology but because expertise and experience are concentrated on these units and because complications are prevented and good medical and nursing care is given. Rothwell has been described as 'the man who stops 10,000 strokes a year' due to his focus on prevention.[33] He and his team prevent strokes – not by using newfangled technology but by conscientiously monitoring the blood pressure of people who have transient ischaemic attacks, or TIAs. Often referred to as 'mini strokes', these are frequently the precursors of much larger, more dangerous strokes.

Despite this, the amount of funding available for the sort of research that Rothwell conducts is dismal. The latest detailed accounts published by the UK's Medical Research Council show that it spent over £814 million on research in the year 2017–18. In that financial year, it spent sixty-nine per cent – over two-thirds – of its budget on basic science research (curiosity-driven research, including animal studies), leaving less than a third for clinical, i.e. human-focused, research.[34] Yet clinical studies are the backbone of medicine, generating the evidence necessary to make good public health, policy and clinical decisions. A 2023 study found, for instance, that patients with type 2 diabetes at a general practice in Southport were able to achieve remission of their disease without drugs, simply by losing weight and following a low-carbohydrate diet. Not only were the patients healthier, but the surgery also only spent £4.94 per patient each year on diabetes drugs compared with the £11.30 per patient spent by neighbouring practices.[35]

In the context of the Covid-19 pandemic, clinical research has been invaluable and scientists have drawn upon a range of different approaches, including clinical trials, human challenge trials – where healthy volunteers are exposed to the SARS-Cov-2 virus in a safe and controlled environment[36] – and 'real world' data.[37] As an example of the latter, Dr Kenneth Baillie and his team at the University of Edinburgh analysed blood samples taken from 471 patients admitted to hospital with Covid-19 in the UK and compared these with samples taken from people with mild versions of the disease, healthy people and those who had previously had swine flu.[38] They found that while several inflammatory proteins were raised in people who were ill, indicating an immune response, only one of the proteins, GM-CSF, was found in severe Covid-19 and was nearly ten times higher in patients who went on to die from the virus.[39] Studies like these, because they are conducted in human populations, are of course immediately relevant, helping to build a clearer picture of how Covid-19 affects humans. In this particular case, it also raises the possibility of developing drugs that can dampen the effect of GM-CSF. Meanwhile, epidemiological studies have been crucial to understanding the prevalence, transmission and progression of Covid-19.

Such relatively low-tech approaches may appear less exciting than the new technologies and methodologies currently being developed, but they are just as important.

Using data obtained from questionnaires, blood samples and biopsies, they are tried and tested, ethical and relatively non-invasive ways of generating data on humans. And of course, they do not involve animals. Another non-invasive and commonly used tool is medical imaging. Magnetic Resonance Imaging (MRI) scans enable scientists to look directly at the various tissues and organs of the human body and produce detailed three-dimensional images that can be used in research. The structure and function of the human brain, or the way it changes with age or disease can, for example, be studied using MRI scans. Post-mortem studies provide another valuable approach; longitudinal studies that include a post-mortem element allow researchers to study an individual with a disease, perhaps one affecting the brain, such as Alzheimer's, both in life and after death, enabling behaviour to be correlated with organic changes and providing a wealth of information on disease processes.

The administration of tiny doses of the psychedelic drug psilocybin to enhance mood and mental health is often referred to as 'microdosing'.[40,41] But microdosing – essentially the administration of a very low dose of a drug to an individual – is also a powerful, non-invasive tool that can be used in human-focused research. In clinical trials, microdosing is sometimes referred to as Phase 0 since it precedes 'first-in-human' or Phase 1 studies. In Phase 0 trials, a new drug is given in a dose small enough to be safe but large enough to allow scientists to study how it behaves in humans, also enabling them to identify drug-to-drug interactions and establish an appropriate dosing level.[42]

A new toolbox

These approaches, in combination with the in vitro and in silico technologies already discussed, mean that we have at our disposal a comprehensive and powerful range of human-relevant approaches. The tools now available for studying and understanding the human body are so much better than the animal methods traditionally used, and the knowledge we are gaining as a result, such as the role that genetics and the microbiome play in health and disease, is astounding. Being directly relevant, such knowledge has staggering potential to advance human medicine.

Having said that, we need to be aware of their potential limitations. Not everyone will be able to afford personalised medicine, or the dietary and lifestyle changes that personalised approaches may require.[43,44] It's also the case that certain things are beneficial for almost everyone's health, regardless of their personal risk status. There is strong evidence that many of the leading risk factors for death and disease worldwide (such as high blood pressure, smoking, outdoor air pollution, alcohol use and a high-salt diet[45]) can be addressed by behaviour change, medication, public health interventions and regulation.[46,47,48] Consequently, personalised approaches may distract attention from existing interventions and policies that are relatively cheap and equitable.[49] And just to put everything into sobering perspective, the main *underlying* causes of ill health are neither behavioural nor medical, but social; poverty, in fact, is the single largest determinant of health.[50]

Personalised approaches need careful thought, then, as does early screening. False positives can be psychologically damaging and may lead to unnecessary treatment. The benefits of screening also need to outweigh any potential harms, meaning that there needs to be an effective treatment available for the condition being screened for. Nevertheless – and bearing in mind these caveats – we can at last dare to imagine the possibility of making real progress using these human-relevant approaches.

Early detection approaches, together with human cell-based technologies, in silico modelling, AI and clinical and epidemiological research, mean we no longer need to conduct research on animals and then translate the findings to humans. Indeed, this now seems a preposterous way to proceed. Why would we use animals when the human genome, for instance, is generating such groundbreaking insights of direct relevance to humans? Why would we use animals when we can now directly investigate humans, using the most powerful and precise technologies, and are able to do this effectively and ethically?

So why *are* we still using animals? What is holding us back?

PART FOUR

THE STRUGGLE TO MOVE FORWARD

11

REGULATORY DYSFUNCTION

A key reason why animal experiments continue is because the regulations governing them are not enforced. UK regulations are supposed to ensure that only essential experiments are conducted; in other words, that research unlikely to benefit humans, or that can be done without animals, is weeded out. However, UK regulators approve and license *every single* application to conduct animal research that they receive. In 2017, the most recent date for which these figures are available, *not one* of the 587 applications to conduct animal research was rejected. Since we know that most animal research does not benefit humans, and that a vast range of animal-free methodologies is available, this looks like an astounding failure of regulation. The situation is similar in most EU member states, but not all; Italy rejected twenty-six per cent of the 1,264 applications it received in 2017, while Hungary rejected thirty-two per cent of its 271 applications and Slovenia fifty-six per cent of its twenty-eight.[1] The EU member states submitting the most applications to conduct animal experiments in 2017 were France (with 3,708 applications) and Germany (3,800), all of which were approved by regulators. Clearly, something is not right.

I spoke to one of the authors of a paper that reviewed these statistics.[2] For almost a decade, Dr Kathrin Herrmann, a veterinarian, was a federal regulator in Germany. During this time, she inspected numerous laboratories and assessed countless applications to conduct animal research. She provides some insight into why so many applications are approved.

'When you receive a well-written research proposal,' she told me, 'as in the scientists know how to make it sound amazing, it's really hard to go against it, there isn't the expertise.'

To properly assess the applications submitted to them, Herrmann tells me, regulators would need knowledge across a wide range of medical fields, a good understanding of experimental design and an awareness of all the non-animal approaches available. But very few people, let alone regulators, whose expertise is normally in veterinary medicine, have such a breadth of experience.

'When you are assessing proposals all day long, trying to meet legal deadlines, you don't have enough time to learn and read up on everything you should know as an inspector. In reality it's hardly ever done, or not done properly.'

Insufficient scrutiny

In the UK, regulators are based in the Home Office's Animals in Science Regulation Unit (ASRU). Here, inspectors are tasked with enforcing the 1986 Animals (Scientific Procedures) Act, generally known as ASPA, which replaced the 1876 Cruelty to Animals Act. ASPA requires UK scientists using laboratory animals to hold a Personal Licence and to work within an institution that holds an Establishment Licence. Additionally, scientists using animals must apply for a project licence each time they wish to conduct a new study. As part of their application, they must show how their research complies with the principles of replacement, reduction and refinement of the use of laboratory animals, usually known simply as the '3Rs', of which more below. They are also required to conduct a harm–benefit analysis[3] by showing how the suffering of animals in their study will be justified by the anticipated benefits of the research. The harm–benefit analysis and the 3Rs together form the cornerstone of UK animal research regulation.

However, ASRU has fewer than twenty-three full-time inspectors and in 2021 these were responsible for checking 137 research facilities and assessing 493 project licence applications.[4] The inspectors simply do not have sufficient time to properly scrutinise all the applications they receive. And because they lack the necessary breadth of expertise to evaluate the need for the research or the likelihood of its success, they draw on their backgrounds as veterinarians when assessing the applications, focusing on the welfare of animals instead of questioning whether the research should go ahead in the first place.

But let's go back a step. Even before scientists reach the stage of applying to the Home Office, they must first obtain funding and then have their research proposal approved by the Animal Welfare and Ethical Review Board (AWERB) within their institution. At each point, proposals are supposedly scrutinised to weed out unnecessary research. Why then, do so many animal studies continue to be funded, and why is every single project licence application in the UK approved?

When funding bodies receive a grant application, they ask experts in the field to assess it, in a process known as peer review. Unfortunately, the expertise of reviewers varies wildly, and while they may be knowledgeable about the topic of the research, they are unlikely to also know about the availability of animal-free approaches in that field, or be able to assess whether the proposed experimental design is appropriate, or whether the benefits of the study are likely to outweigh its harms. Consequently, while funders expect peer reviewers to scrutinise these aspects, many reviewers simply lack the expertise to do so, instead accepting at face value applicants' claims about the necessity of using animals and the anticipated benefits of their research.

Unfortunately, a similar lack of expertise is to be found on AWERBs.[5] These boards should be composed of scientists, veterinarians, animal care staff and lay members. Some, but not all, include statisticians so there is frequently insufficient oversight of experimental design,[6] meaning that good science (and therefore useful findings) cannot be guaranteed. There is also no requirement to include an expert in animal-free approaches, making it impossible to assess whether these could be employed instead of animal models. As for lay members, they are usually recruited from the board's institution so are likely to share its institutional values, including a belief in the importance of animal research.[7,8] As I have already mentioned, when I attempted to join one of these boards as a lay person, the chair made it clear that he was only looking for members who were fully signed up to the animal research enterprise. It's hard, then, not to regard AWERBs as rubber stamping committees rather than genuine opportunities for scrutinising whether or not an animal study should proceed.

In reality, ethical review boards are dominated by scientists and it is their interests that take precedence.[7] For many years, Professor John Gluck was chair of the US equivalent of AWERBs at his university. He recalls that scientists were very resistant to ethical review when the process was first instigated and highly indignant about having to justify their research to others. 'The level of hostility and condescension I encountered was at times incredible', he writes. 'Some researchers accused the committee of having no members intellectually capable of understanding the purpose of their methods and design.'[9]

If this was the experience of the chair, who at the time was an established scientist and animal researcher, one can only imagine how difficult it is for those further down the pecking order to challenge an application. If a scientist simply declares that their study will help sick children – as was the response to a lay member's question in an AWERB meeting I observed – then the issue is unlikely to be pursued. Non-scientific members of the boards usually lack the necessary confidence and authority to provide effective scrutiny.

The evidence suggests – and my experience of observing a meeting confirms – that the focus in AWERBs is on harms to animals, with very little consideration of the likely benefits of the research.[5] This may be partly because harms are easier to assess than benefits, which rarely materialise until some point in the future (if at all),[10,11] and partly because the benefits are simply assumed. There may also be a perception that the research must be worthwhile because it has been funded. Consequently, rather than deciding whether or not the research should proceed, the assumption among board members – just as with Home Office inspectors – is that it *will* go ahead and that their only role is to minimise animals' suffering. Of course it is essential to reduce the suffering of animals, but if a rigorous harm–benefit analysis were conducted in the first place, most studies would not gain approval[12] and animal use would be completely avoided.

In 2003, Dr Dan Lyons, now CEO of the Centre for Animals and Social Justice, won the legal right to obtain and publish confidential documents that gave detailed descriptions of UK experiments involving the transplantation of genetically modified pig organs into non-human primates. The experiments had been conducted by pharmaceutical company Imutran, in the hope of demonstrating that a pig heart or kidney could support life in a non-human primate, supposedly paving the way for human trials. The documents reveal that the Home Office accepted Imutran's rating of their research as 'moderately severe', even though the experiments left many animals dead or in extreme suffering,[13] and that, over several years, it repeatedly approved the research on the basis of the company's claims. Lyons concludes that the Home Office failed to conduct its own harm–benefit analysis and had a cooperative relationship with the scientists, who were the dominant party. This seems to be a case of 'regulatory capture', whereby regulators become sympathetic to, or led by, those they are supposed to be regulating.

For as long as regulators and ethical review boards

continue to facilitate rather than properly scrutinise project licence applications, then, animal experiments will continue.

Encouraging the continuation of animal research: the 3Rs

In 1959, zoologist Bill Russell and microbiologist Rex Burch published *The Principles of Humane Experimental Technique*,[14] a landmark book in which they outlined the principles of replacing animals with non-animal approaches wherever possible, reducing the number of animals used to the absolute minimum necessary and refining experimental procedures as far as possible to minimise animal suffering.[15] The 3Rs are now firmly embedded in the policies of organisations that fund, promote or conduct animal experiments.

Refinement, which, unlike the other two principles, does not challenge the ongoing practice of animal research, receives the lion's share of attention. Even so, numerous studies have shown that scientists pay only lip service to this principle, frequently failing to take opportunities to reduce animal suffering.[9,12,16–21] The regulations themselves convey a mixed message, specifying on the one hand that suffering should be minimised but on the other, permitting severe harms to be inflicted. In 2021, 149,917 experimental procedures, almost eight per cent of the UK's total, were classified as severe,[22] and severe harms are also permitted within EU and US legal and regulatory frameworks.[9] In the UK, there are fears that changes in the way ASRU inspects animal research facilities will lead to even fewer in-person inspections than previously, removing a further layer of animal protection.[23]

As for the reduction principle, this is virtually ignored,

with animal use continuing unabated and even increasing.[24] Over nine million laboratory animals are used annually in the EU alone,[25] and Humane Society International estimates that more than 115 million animals are used globally in research each year. Another estimate of annual worldwide use is even higher, at 192 million.[26] The latest US statistics (which do not even count mice, rats or birds[27]) show an increase of almost six per cent,[28] while in the UK there was also a rise of six per cent, according to the most recent figures.[22] And although there are plans in the US and the Netherlands to reduce the number of animals used in regulatory safety testing, these have faced setbacks, as we'll see in the last chapter. Meanwhile, there are no plans for reducing the number of animals used in basic and applied scientific research, which together account for the bulk of animal use.

Replacement is faring no better. In 2013, UK animal research regulations were aligned with European Directive 2010/63/EU[29] which emphasises the principle of replacement, stating that 'the use of animals for scientific or educational purposes should ... only be considered where a non-animal alternative is unavailable'.[29] This directive was retained in UK law after the UK's departure from the EU,[30] but the replacement principle is enforced by neither funding bodies, AWERBs nor Home Office regulators. In a parliamentary debate which took place in January 2023 about commercial breeding for animal laboratories, Sarah Dines, Parliamentary Under Secretary of State at the Home Office, declared, 'Robust regulation will ensure that animals are not used where a non-animal alternative could deliver the benefit sought.'[31] But if the regulations were robust, surely at least a few of the proposals submitted to the Home Office would be rejected?

In reality, the regulations are far from tough. Researchers are not required to prove that they have seriously investigated the possibility of conducting their research without animals; all they have to state in their project licence applications is that they have considered this.[9] Analysis by Animal Free Research UK found that scientists frequently do little more than list a few non-animal methods and claim that these are unsuitable, rather than showing evidence of any serious effort to identify animal-free approaches.[32] There is no expectation that they will be challenged about this, and even if they were, it would be too late; by the time a scientist applies for a project licence they have usually already acquired funding which was granted on the basis of conducting animal experiments.

Proper enforcement of the replacement principle requires intervention earlier, when scientists are planning and seeking funding for their studies. At this point, things are still sufficiently fluid to enable a change of direction. But by the time a research proposal has been approved by an AWERB and reached the Home Office, it is virtually set in stone; to imagine that at this stage the methodology might be overturned and replaced with a non-animal approach is frankly absurd. The replacement principle as it currently functions, then, is no more than a symbolic box-ticking exercise.

In 2015, hopes were raised when a roadmap for replacing laboratory animals with non-animal technologies was published by Innovate UK, the government's innovation agency.[33] Unfortunately, however, no action was taken and the initiative stalled.[30] So, we find ourselves in a situation where, in 2020, 452 skin sensitisation tests were conducted on mice in the UK, despite the availability of a validated method that does not use animals. Meanwhile, it is estimated that each year, up to three-quarters of a million animal tests take place in the EU for which validated non-animal tests are available.[32] So much for the 3Rs. While they may function to relieve or avoid some animal suffering, they are failing dismally to reduce or replace animal use. Consequently, there are increasing calls for the 3Rs to be replaced by the 'one R' of complete replacement. While animal research continues, it is of course essential to ensure that suffering is reduced, but suffering could be completely avoided if the replacement principle were properly enforced.

In taking animal research for granted and reinforcing a perception of it as standard practice, the 3Rs divert attention away from the need for a science based on human biology. And they have another consequence. Whenever objections to animal experiments are raised, scientists and regulators appeal to the 3Rs as a quick means of providing reassurance that our moral obligations to animals are being managed.[34] Indeed in 2004, one of the ways in which the UK government responded to mounting public opposition to animal research was the creation of the National Centre for the Replacement, Refinement and Reduction of Animals in Research, commonly known as NC3Rs. In promoting an *appearance* of regulation, then, the 3Rs function to allay the public's concerns, maintain the status quo and perpetuate the use of animals in science. Michael Balls, Emeritus Professor of Cell Biology at the University of Nottingham and former head of the Fund for the Replacement of Animals in Medical

Experiments, was for many years a staunch advocate of the 3Rs. Now retired, he acknowledges that many scientists are not truly committed to the principles and that one of their key functions is to deflect criticism. More attention should be paid to the scientific justification for conducting experiments on animals to begin with, he suggests.[35,36]

In 2017, at the tenth World Congress on Alternatives and Animal Use in the Life Sciences, which was held in Seattle, I attended a talk given by a veterinary specialist who has played a key role in UK animal research regulation over the years. She spoke at such length about the need to sustain public confidence in the regulations that a member of the audience eventually leapt to his feet in frustration.

'Is the function of the regulations simply to make the public believe that things are done properly?,' he asked furiously. 'If only they knew!'

The regulations have two functions, then. One is to enforce the law governing animal research, and the other is to reassure an uneasy public. UK regulators do better at the second than the first. The very existence of the regulations reassures many of us that animal use is being carefully scrutinised, putting our minds at rest. We may feel uncomfortable about what goes on behind the closed doors of laboratories, but we are assured that the use of animals in science is well regulated and regularly reminded that UK regulations are among the strictest in the world. So we are soothed and quietened, and the use of animal research continues.

What would happen if the regulations governing animal research were properly applied? When we were both at the University of Bristol, Professor Christine Nicol and I conducted a retrospective harm–benefit analysis of over 200 animal studies conducted worldwide. It was the first time such a study had been conducted; the harm–benefit analysis conducted for regulatory purposes is *pro*spective, so it only considers *potential* harms and benefits. But after weighing the *actual* harms to animals against the *actual* benefits to humans and taking into account the scientific quality of the studies, we concluded that fewer than seven per cent of the studies should have been permitted.[12] Had we also considered the availability of suitable non-animal approaches, we would have likely reduced this proportion still further.

Regulating medicines

In 2019, Dr Mihael Polymeropoulos, CEO of pharmaceutical company Vanda, took the FDA to court after one of its divisions, the Center for Drug Evaluation and Research, demanded the company conduct a nine-month study on dogs prior to testing its experimental drug tradipitant in humans. As tradipitant had already undergone extensive preclinical testing, including in animals, Vanda challenged the request, stating that additional dog studies would not provide any further useful information. The FDA responded by preventing Vanda from proceeding to human trials, thereby thwarting any further development of the drug. In response, Vanda filed a lawsuit against the FDA, charging it with 'arbitrary and capricious behaviour' in requiring the dog studies. Polymeropoulos stated that there was no evidence the dog studies would add value, given the drug's well-established safety profile in animals and humans, and he noted that the FDA had provided no scientific justification for requesting them.[37]

'It is time to demand that the pharmaceutical industry and government regulators abandon unscientific, low-resolution animal testing,' he declared, 'and adopt modern, human-based scientific methods to advance human drug safety.'[38]

Polymeropoulos won an award for his courageous efforts to modernise drug testing, but in January 2020, a judge ruled against the company. Vanda has been able to pursue the development of tradipitant for short-term use, but at the time of writing, the FDA is still preventing the company from developing the drug for use beyond twelve weeks unless long-term studies in animals are conducted.[39]

As the Vanda case shows, then, another way that animal use is perpetuated is via the regulations governing the safety testing of medicines. Prior to being trialled in humans, new drugs are tested for safety by pharmaceutical companies whose scientists may use a combination of animal tests, in vitro cell studies and in silico modelling to generate the necessary information on dosing and toxicity. Pharmaceutical companies are guided by national regulatory agencies such as the FDA or the UK's MHRA. If a new drug candidate is judged to be safe in preclinical tests and subsequent clinical trials, the national regulatory agency will grant approval for the company to market it.

However, regulatory agencies tend to regard animal tests as the 'go-to' preclinical tests because they have been used for decades and the whole international regulatory system is attuned to them. Regulators have become confident about using animal data as the basis for making their decisions about drug safety and, conversely, are unfamiliar with the potential of human biology-based methodologies. They are also understandably cautious since human lives are at stake and may therefore hesitate to base their decision-making on data generated from new methods.

The situation is not helped by disagreement about how to demonstrate that new methods are reliable and valid.[40] Some feel that the bar is set too high for validating human-focused technologies, with the quest for perfection causing delays and bottlenecks. Others believe that new approaches should be validated against animal data, which is obviously problematic since animal methods themselves have never been validated[41] and are far from the gold standard, as we know. Furthermore, data from human biology-based methods would not be expected to agree with data generated from animal studies. Fortunately, there seems to be a growing consensus that the rational way to proceed is simply to assess new technologies in terms of their ability to provide data that are relevant to humans and that will give regulators confidence about the likely human response.[42]

'The regulators are scared of approving something that's going to harm people,' says Dr Jarrod Bailey, Science Director of the charity Animal Free Research UK, 'but they don't realise that their way of doing things is not scientific. Once they realise there are better ways of protecting people, then we'll get somewhere. They need the confidence to make the change.'

Regulators might feel more confident if they had greater familiarity with new approaches, but unfortunately, although many major pharmaceutical companies routinely use human in vitro tests to screen and select new drug compounds 'in-house', they do not always include data generated from these tests in their submissions to regulators. Consequently, regulators are not accustomed to seeing data from human biology-based tests and have therefore been unable to gain an appreciation of their value in the context of safety testing. In turn, this means they are unable to provide clear guidance to other scientists who might want to use human-focused methods. Lacking clear guidance from regulators about the acceptability of new approaches, these scientists worry that their human-based data could be rejected by regulators, resulting in delay and lost revenue.[43] So, in a vicious cycle, they continue to submit animal data in the belief that it will be the simplest and fastest route to gaining regulatory approval.

Again, due to a lack of confidence in the response of regulators – although there is now an option in some cases to use a single species in the preclinical testing of conventional drugs,[44] rather than conducting tests in both a rodent and a non-rodent species – many companies continue to use two species in a 'belt and braces' approach, worrying that otherwise regulators will refuse their applications or question the robustness of their evidence.[45]

There is an obvious need, then, for regulators to spell out their requirements much more clearly. At present, although they require assurance that a particular substance will not cause harm, regulators do not necessarily mandate how that assurance should be provided.[46] Both the MHRA and an expert in EU and international animal law have stated that there is actually no requirement to use animals[30] in this context, and the MHRA claims to encourage the use of in vitro methods for testing new drugs. Nevertheless, scientists from UK biotech companies point out that the regulator is giving mixed messages, since the wording in the MHRA guidance refers almost exclusively to animal use for preclinical testing, clearly driving an expectation that animal tests are required.[30]

This lack of clarity is thwarting progress. But there is good news. As 2022 drew to a close, President Joe Biden, in a highly significant development, signed into law the long-awaited FDA Modernization Act. This landmark legislation amends the wording of the 1938 Federal Food, Drug, and Cosmetic Act to clarify that non-animal approaches may be used to investigate the safety and effectiveness of drugs.[47] As such, US scientists will now be able to choose the most appropriate method to discover and develop drugs rather than being forced to rely on animal models. Furthermore, although the Center for Drug Evaluation and Research within the FDA tends to rely on animal data, as the Vanda example shows, a cross-FDA working party is proactively investigating the potential of new methodologies; it has established an Alternative Methods Working Group to explore how human-focused approaches might improve the regulator's ability to bring effective drugs to market faster, while at the same time preventing unsafe drugs from reaching the market. It has also instigated a webinar series to give developers of new technologies an opportunity to showcase their methods to FDA scientists.[48]

In the UK, where twenty-one per cent of all experimental procedures conducted on animals in 2021 were to satisfy regulatory requirements for safety testing,[22] we can only hope that the US example will encourage the MHRA to modernise our regulations similarly and to proactively investigate the potential of new approaches.

It's important to note, however, that not all responsibility for the continued use of animals by pharmaceutical companies can be laid at the door of regulators. Companies may benefit from regulations that mandate the use of animals in safety testing because, at present, this gives them legal protection if they are taken to court by individuals who have suffered adverse drug reactions. Put simply, if a pharmaceutical company faces litigation for damages, it can say that it tested the drug on animals and the tests showed the drug was safe.[46] After all, there will always be some animals that will give companies the answers they're looking for. This is another reason why the regulations need to be urgently reformed; once animal data are no longer considered the gold standard, companies will be unable to protect themselves in this way.

Thwarting transition

In the US at least, then, progress is being made in terms of the regulations governing pharmaceutical drugs. No such progress, however, is evident in relation to the regulations governing animal research. Here, insufficient expertise, inadequate scrutiny, scientific self-interest and regulatory capture thwart the proper implementation of Directive 2010/63/EU which, if correctly applied, would massively reduce the number of animal studies conducted and see the vast majority of experiments being replaced by human-focused approaches. These regulations are in need of a complete and utter overhaul, yet the *appearance* of regulation lulls us into a belief that all is well, subduing dissent and preserving the status quo.

Regulations play a key role in thwarting the transition to human-focused approaches, then, but they are not the only factor. Additional forces are at work.

12

LOCKED IN

As a schoolboy Geoff Pilkington enjoyed biology, but in general he preferred cricket, rugby and athletics to academic subjects. Nevertheless, after leaving school in 1968, he secured a job as a technician in a research laboratory at the Royal Free Hospital, which was then based in Hunter Street, central London. He recalls that, at the time, practically all the research involved animals, mostly rats and mice but also cats, dogs and non-human primates. Moving on to the National Hospital for Nervous Diseases and then the Middlesex Hospital Medical School, he developed an interest in brain tumour research and went on to conduct his own investigations, attempting to induce brain tumours in rats and mice using chemical carcinogens and gaining a PhD in the process. But he began to realise there was a problem when it came to translating his findings to humans.

'The results were at best misleading, because if you looked at the counterpart situation in humans, you found that protein expression differed, and of course the growth factors differ, and there's a whole load of things – including different blood vessel structures, differences in pattern of tumour cell invasion and types of cell involved – that rendered this research not only inhumane but also really lacking in relevance and scientific rigour.'

In 1979, in a horrible twist of fate, Geoff Pilkington's mother was diagnosed with a malignant brain tumour and died only sixteen months later. This only increased his motivation to develop an approach that would deliver for patients. In a new post at the Institute of Psychiatry in South London, he began to experiment with tissue culture, starting with animal tissue but soon moving to human tissue when he realised he could source human cells from the Institute's neurosurgical facility. Over time, he built up a team of researchers and established a successful animal-free laboratory, developing complex, dynamic, human biology-based models that reflected the human brain tumour state. I remark that his transition from animal to human-relevant research sounded quite smooth.

'No, not at all,' he says, 'because you were always under pressure to do animal work. The animal work was still the norm, and so it was seen as a little bit quirky, having a non-animal lab, a human-based lab, taking human tissues in and creating cultures and doing tests on those. And there was a thing called the Research Defence Society. At one point, its director invited me to his London office and tried to persuade me of the value of animal research, but I just carried on developing the human research programme on brain tumours.'

'Given that it would have been a lot easier to stay with animal research, what was it that made you swap then?'

'I think it was really the science. I thought, well, if you look at the science, we're actually using animals for no net purpose. The goal was to get medications and approaches for curing patients, and this was not fit for purpose.'

Securing funding for his work was a constant challenge, and at first, Pilkington was only able to obtain small grants from a handful of charities, including those that specifically funded animal-free research. When he approached the larger funding bodies, he was often asked to expand his proposal to include an element for confirming his findings in animals. And he frequently met with a similar response when submitting his findings for publication in scientific journals.

'I had this really rather innovative approach of having an all-human lab,' he tells me, 'but they still kept coming back to us, whenever we submitted a grant application or a paper, asking, "what animal model are you going to use to validate that?".'

In 2003, he moved his laboratory to the University of Portsmouth where he became Professor of Cellular and Molecular Neuro-oncology and headed up the Portsmouth Brain Tumour Research Centre until his retirement in 2019. There, his team of some twenty-four researchers developed a range of human-relevant in vitro models for studying brain tumours and delivering drugs to the brain. Does he think things have improved since he started out in the 1960s?

'In some ways, yes. There is now substantially more research carried out without using animals, more young researchers forging their careers in an animal-free environment and many animal species no longer seen in the laboratory. But there are still these dyed-in-the-wool folks, so we've got a battle to fight against that part of the scientific community who are still embroiled in using animal models as the be all and end all. While we still have grant-giving bodies saying you must use this or that animal model, it's going to be perpetuated.'

Funding streams lock scientists out of human-focused research

Even now, scientists conducting human biology-based research, like Geoff Pilkington, find it difficult to secure funding for their work. When competing with researchers using established, animal-based approaches, they are at a disadvantage due to a lack of awareness of innovative methodologies and because these non-animal approaches have fewer dedicated funding pots. The main UK funder is the NC3Rs, which in 2019 provided an estimated £2 million for research on new approaches.[1] This is a miniscule amount. Funding for basic science human biology-based approaches is estimated to represent between 0.2 per cent and 0.6 per cent of the total biomedical research funding in the UK and around 0.02 per cent of the total public expenditure on research and development, which for 2019–20 was £10.45 billion.[1] In 2022, the Biotechnology and Biological Sciences Research Council earmarked £4 million 'to drive the development and uptake of non-animal technologies for bioscience research', but from a total budget of £1,243.06 million, this is a trifling amount.[2] The situation leaves many scientists scrambling for philanthropic donations or applying to small charities. Animal Free Research UK and the Humane Research Trust are the main charities in this arena and, over the decades, have distributed several million pounds in grant funding since they were established over fifty and sixty years ago respectively, each typically awarding several hundred thousand pounds annually.[3,4]

A similar pattern is found in the US where, for example, in 2019, the National Institutes of Health provided almost $11 million for animal research into breast cancer, but less than $2 million for breast cancer research using human biology-based methods.[5] The largest public funder of biomedical research in the world, the National Institutes of Health currently has no programme dedicated to these innovative approaches.[6]

Institutions lock scientists into animal research

At the beginning of this book, and drawing upon the insights of economist Joshua Frank,[7] I explored how, over the course of the twentieth century, animal research became locked into institutions through a variety of positive feedback mechanisms. Suffice to say, these mechanisms have only solidified over subsequent decades, presenting a real obstacle to progress. Consequently, even if they secure funding, scientists using in vitro or in silico approaches face further challenges. Geoff Pilkington's experiences continue to be worryingly common, with a 2022 survey finding that a third of those questioned had been asked by journal reviewers to add animal data to their non-animal study.[8] The persistence of this inappropriate view of animal studies as the gold standard makes it difficult for scientists using human-based methodologies to publish their findings unless they include an element of animal research. Indeed, journal editors' requests for animal studies are so frequent that some researchers conduct animal experiments alongside their in vitro or in silico experiments simply to give themselves a better chance of having their findings published.[8,9] What can explain this mindset, in which animal models continue to dominate?

Some scientists resist a transition to human-relevant approaches because they retain confidence in animal models, while others believe that it is still too soon to relinquish animal use. But it is also the case that scientists in general tend to be resistant to change. The philosopher Thomas Kuhn pointed out that while students of music, art and literature are exposed to the ideas and work of other musicians, artists and writers, students of the natural sciences rely mainly on textbooks for their education. The 'scientific training is not well designed to produce the man who will easily discover a fresh approach', he wrote, also referring to it as a 'narrow and rigid education, probably more so than any other except perhaps in orthodox theology'.[10]

Many believe scientists to be open-minded, critical thinkers who put their trust in evidence, but the scientific education can create a pressure to conform, leading scientists to be sceptical of ideas that challenge the orthodoxy. If this is the case for scientists in general, those using animal models are likely to be even less receptive to new ideas because, for over a century, they have insulated themselves from society at large and even from other scientific disciplines. Closing ranks to protect themselves from protestors and speaking only to other animal researchers who confirm and reinforce their beliefs, the result is a 'pathological consensus'.[11] These scientists may also be psychologically locked into the practice, unable or unwilling to consider the possibility that animal models are unproductive and that the suffering they inflict may be in vain. Defending their work in the face of ethical objections may have the effect of shoring up their beliefs still further.[7]

There are, of course, exceptions – scientists like Geoff Pilkington who begin their careers using animals but who become curious about the potential of human biology-based approaches. Academic culture, however, does not make it easy for such scientists to switch, especially if they have been using animals for a while. Within academia a researcher usually has to specialise in a single subject or methodology if they are to have a successful career. After a few years of conducting and publishing research, they come to be regarded as experts in their field and are invited to write papers, give keynote speeches at conferences, edit journals and become fellows of this or that society. As they rise up the academic career ladder, then, they become more and more locked into their topic. For this reason, a scientist who uses animal models is likely to continue using animal models. They may even frame all their research questions around one particular animal model.[12] Having spent years developing their laboratories and research teams based on animal models, and with all their expertise, funding and publications dependent upon animal use, a transition to human-based approaches represents a significant risk, potentially jeopardising their career and funding opportunities.[13] Consequently, even if a scientist wanted to switch to human biology-based methods, they would find it very challenging to do so in the absence of specific support and funding. This goes some way to explaining why such scientists frequently conduct human biology-based research *in tandem with*, rather than instead of, animal studies.[14]

Research suggests that an innovation is more likely to be adopted within an organisation that is receptive to change and new knowledge, and if the innovation fits with the organisation's existing values, norms, technologies and systems.[15] But the systems, values and norms within many universities revolve around animal use and there is no perceived need for a new approach. Universities are monolithic, inflexible entities. They resist change and are slow to adapt. Some may even repel innovation.

From 2005 until 2017, Professor Merel Ritskes-Hoitinga, whom we met in an earlier chapter, was manager of the animal research facility at Radboud University in the Netherlands. There, she also established and ran a successful centre dedicated to advancing the use of systematic reviews of animal research. All was going well until she began to voice her doubts about the value of animal experiments and to talk about the potential of human-based approaches. Then, she was advised that her role was to provide a service to users, not comment on the science. Things came to a head in 2013, following an interview she gave to the Dutch newspaper *Trouw*, in which she stated her opinion that animal testing could be reduced by eighty per cent. This sent shock waves throughout the Dutch animal science community and two of her colleagues openly disagreed with her in a letter to the newspaper. The research directors at her university began to withdraw their support and her working life gradually became more and more uncomfortable. Eventually, she relinquished her post as manager of the facility, taking up a different position within the university in 2017 and finally leaving Radboud altogether in 2022.[16]

It's perhaps not surprising that Ritskes-Hoitinga's stance met with such opposition; in questioning the very purpose of the facility she was managing, she was in effect threatening the interests of those who depend upon the facility for their research, careers, funding and publications. As we saw in Chapter 2, rodents can be transformed easily into publications,[17,18] and publications bring funding to universities. Although many imagine universities to be liberal, open-minded institutions, these qualities only go so far when funding is at stake. But not all universities are the same. In 2022, Ritskes-Hoitinga joined Utrecht University as Professor in Evidence-Based Transition to Animal-Free Innovations. Utrecht University is at the forefront of a move to animal-free, human-based approaches in the Netherlands, a move it regards as crucial to the improvement of scientific research.

Unfortunately, Utrecht is an outlier at the moment. The dominant mindset within academia is that animal models can be improved and will ultimately 'come good'. So scientists expend considerable effort and resources producing all sorts of checklists and guidelines in an attempt to fine-tune their animal studies or improve the way they are reported, rather than relinquishing them for something better.[19,14] These endeavours are futile, however, because tinkering with the technical aspects of animal models does nothing to address the paradigm's fatal flaw, which is species differences.[20]

Thomas Kuhn reminds us that throughout history, when confronted with anomalies that challenge a paradigm, scientists have tended to retain a strong belief that the paradigm will ultimately solve all its problems. Consequently, instead of renouncing it, they attempt to modify it and eliminate any apparent conflict.[10] The animal research 'improvement project' can be understood as just such an attempt to modify the paradigm and is typical of the sort of response that occurs when a scientific paradigm enters a period of crisis. But Kuhn also noted that for as long as anomalies are regarded as superficial, there will be no revolution; to provoke radical change, an anomaly has to reveal real inadequacy in a paradigm. This means that, while the challenges of translating findings from animals to humans continue to be perceived as merely technical and resolvable, there will be no scientific revolution. Change will only occur when it is acknowledged that the very foundation upon which animal research is based is defective.

Vested interests cement lock-in to animal research

In an earlier chapter we saw how, from its outset, multiple industries sprang up to supply and service the animal experimentation enterprise. In the intervening years, these industries have flourished and prospered, and the use of animals in science has become a massive global business with numerous players, including specialist breeding companies, suppliers of surgical equipment, cages, restraints and animal feed, as well as testing laboratories and pharmaceutical companies.[21] The pharmaceutical industry is one of the most profitable industries in the world and animal use has played a key role in its success. Indeed, although projections vary, the animal testing market is predicted to expand over the coming years,[22] the reason being increased demand and 'preference' for animal models.[23]

'In short', write philosophers Hugh LaFollette and Niall Shanks, 'animal experimentation has spawned, and is supported by, a complex web of economic relationships. Many people earn their livelihood, directly or indirectly, from the practice of animal experimentation. These people will be disinclined to criticize or radically alter the practice since so much is at stake economically. Although this does not imply that animal experimentation is simply done for economic reasons, we think that it would be folly to ignore the ways that economic considerations make a candid appraisal of the practice more difficult.'[24]

Economic interests alone, then, do not explain the continued use of animals in science, but they are highly significant, and stakeholders are fighting hard to protect them. As American author Upton Sinclair apparently commented, 'It is difficult to get a man to understand something when his salary depends upon his not understanding it.'[25]

Cultural mindset

What about us? Are we, the general public, also partly responsible for the continued use of animals in science? Perhaps we are too complacent, too trusting when scientists tell us that animal research remains essential. Maybe we find the topic too uncomfortable to think about and wilfully keep ourselves in a state of ignorance.[26] If we knew more, would we advocate for change?

A cultural mindset in favour of animal research seems to exist, best expressed in the phrase 'necessary evil', and a deep (if unacknowledged) cultural belief in its value appears to have taken root, hindering progress. This is not altogether surprising, given that for decades, scientists and the media have been singing its praises. Biologist Lynda Birke and colleagues suggest that, as a culture, we cast laboratory animals as our saviours, the creatures who bear our suffering for us.[17] After all, the deaths of laboratory animals are frequently referred to as 'sacrifice',[17,27] the implication being that they die for our sake.[28] Some anthropologists see sacrifice as a 'ritual of defence' against our human vulnerability to unpredictable and uncontrollable events,[29] while others observe, interestingly, that a sacrifice doesn't necessarily have to achieve a successful outcome to be considered effective. Because the value is mostly symbolic, its outcome is not the most important issue.[30] On a symbolic level, then, perhaps the daily ritualistic sacrifice of laboratory animals somehow reassures us that something is being done to protect us from illness and death.

False reassurance

But we need to rouse ourselves from our slumber and realise that this is false comfort. Twenty years after the death of Geoff Pilkington's mother from a brain tumour, my outrageous, funny and talented younger brother Caspar, a successful record producer, was also diagnosed with a brain tumour. Although he underwent surgery for his oligodendroglioma, the surgeons were unable to remove it all. There was no effective treatment available to him. Chemotherapy and radiation would only have postponed the inevitable and would probably have harmed him. He was only thirty-three years old when he died in 2004. Almost twenty years on, and despite decades of animal research, the options remain the same for people with brain tumours: surgery, radiation and chemotherapy or, as Professor Azra Raza puts it, 'slash, burn and poison'.

How much longer should we be expected to put up with this state of affairs, to watch our loved ones suffer and die? And for how long should we tolerate such poor returns on our investment in research? It is we, after all, who fund research through our taxes and charitable donations. Who, if anyone, is responsible for policing science? Some say we are all responsible[31] but in practice this means that none of us is.[11] There has long been a perception that science is self-correcting, but this view is increasingly challenged,[32,33] and rightly so. The failure of the animal research paradigm can no longer be left to scientists to fix. There is too much at stake.

13

DEATH THROES AND BIRTH PANGS

At the beginning of this book, we saw that the practice of animal experimentation managed to gain a foothold during the nineteenth century, not because it led to advances in human medicine but because the experimental method had become associated with 'proper science'. Conducting animal experiments therefore helped elevate the formerly lowly professions of physiology and medicine, giving them a new respectability. The narrative of medical progress merely served to justify these experiments to a shocked public, enabling these professions to flourish. Over the course of the following century, big business and various lock-in mechanisms combined to ensure that animal use – and in particular 'animal models' of disease – became firmly embedded within biomedical research. Yet, we have learnt that the scientific quality of most animal studies is startlingly poor, resulting in untrustworthy findings and entire bodies of research that are so biased that they appear more positive than they actually are.

Perhaps unsurprisingly, then, we have seen that clinicians

and clinical researchers tend to draw upon human rather than animal data when investigating disease or developing guidelines for human medicine. And the evidence shows that, compared with the impact of prevention, public health and clinical medicine, the contribution of animal studies to human health has been negligible. That some of our most common diseases continue to lack treatments despite a century and a half of animal research speaks for itself.

This book has argued that a transition from animal research to human biology-based approaches will make drug testing more reliable, with the upshot that new drugs will be safer for humans, with fewer unforeseen adverse effects. Furthermore, without unreliable and misleading animal models holding us back, human-focused medicine will finally be able to realise its potential; new treatment approaches will be developed and there will be a greater focus on detecting the earliest signs of disease to enable timely intervention. In short, lives will be saved.

Detractors will point to cases where animal research has produced useful knowledge or led to medical advances. I don't seek to deny this; my argument is that we should abandon animal research because it is *unreliable*. There is even a respected theory – evolutionary theory – that explains why this is the case. Conversely, however, there is no robust explanatory theory underlying the use of animal models. Instead, the paradigm relies on a hypothesis based on homology, a hypothesis that has been repeatedly tested and disproven. So animal models may occasionally be productive, but they will never be consistently and reliably productive because of species differences. It's hit and miss. And we can do better than hit and miss. Species differences undermine the whole premise of animal research that is intended to have relevance to humans. It is the paradigm's fatal flaw, not an issue that can be overcome. But until this is properly acknowledged, there will be no revolution.

Thankfully, more and more scientists, policymakers and politicians are recognising that human biology-based approaches are the way forward and, increasingly, efforts are being directed at transitioning away from animal research. In this final chapter, we explore some of these exciting developments but also the considerable pushback from animal researchers and others with vested interests.

Reasons to be cheerful

In 2019, the Wellcome Trust's Sanger Institute in Cambridgeshire, a world-leading genetics laboratory that breeds mice, rats and zebrafish for medical experiments, announced that it would close its animal facility within the next three years. The reason?

'Our science strategy is changing. It is as simple as that,' said Sir Jeremy Farrar, then Director of the Wellcome Trust.[1] 'New laboratory techniques have recently been developed which mean we simply do not need the numbers of animals that were once required for our experiments.'

There was opposition from some scientists, who felt that the movement away from animals was premature, but Farrar insisted that this was a good news story, noting that the development of tissue culture and organoids had opened up the possibility of reducing animal experiments. Professor Mike Stratton, Director of the Sanger Institute, agreed.

'This has been a difficult decision,' he said, 'however, we believe it is the best way to continue to deliver science and make discoveries that impact on human health.'[1]

In the same year, the Medical Research Council announced that research at its Mammalian Genetics Unit at the Harwell Institute in Oxfordshire, the UK's leading centre for mouse genetics, was going to be discontinued in a move the Council described as 'a reflection of the changing scientific landscape'.[2]

These encouraging developments had been preceded a few years earlier by a flurry of reports from the Home Office and government agencies acknowledging that breakthroughs in human biology-based technologies were creating opportunities for the UK to reduce its reliance on animal research.[3,4,5] Meanwhile, the Dutch government launched its 'Transition Programme for Innovation without the use of animals' in 2016, declaring its aim to eliminate animal use in safety testing by 2025. And in 2021, the European Parliament voted by a decisive 667:4 to ask the European Commission to establish an EU-wide action plan to actively phase out the use of animals in experiments. Shortly after this, the European drugs regulator, the European Medicines Agency, launched the Innovation Task Force to work towards replacing laboratory animals with innovative technologies in the context of drug testing.

Across the Atlantic, the first signs of progress had come earlier, beginning in 2007 with the publication of a landmark report. *Toxicity Testing in the 21st Century* laid out the steps necessary to transition from using animals in safety testing to a new paradigm 'firmly based on human biology'.[6]

'I remember reading it,' says Kathy Archibald, Founder of Safer Medicines Trust, 'I stayed up late into the night reading it, I was so excited. I thought, *this is the end*. They predicted there would be major change within ten years and a complete end of the practice within twenty.'

The report's authors argued that a paradigm based on human biology would be more relevant, lead to huge cost and time savings, and markedly reduce animal use. Furthermore, they noted that all the tools necessary to achieve this shift were either currently available or in the advanced stages of development. This pivotal report spawned a number of similar initiatives by US federal agencies, including the FDA,[7] the Environmental Protection Agency (EPA) and a committee of sixteen federal government agencies,[8,9] all aiming to replace animals with human-based approaches in safety testing. Excitingly, in 2019, the EPA announced its intention to eliminate testing on mammals by 2035 and to focus instead on in silico modelling and new technologies such as organs-on-chips.[10] And, as I have already mentioned, the 2022 FDA Modernization Act clarifies that US drug manufacturers can use a range of different technologies to investigate drug safety and efficacy, rather than just animals.

So there are many reasons for optimism. Dare we begin to envisage a world without animal research, and if so, what might this look like? Well, in addition to improving public health, a transition to human-relevant technologies will improve the public purse. At present, the average cost of developing a successful new drug, factoring in the cost of all the failures, is around $2.6 billion,[11] with each new drug taking up to ten years to develop.[9] If we use technologies that more accurately predict human outcomes, fewer new drugs will fail, saving money and reducing the cost of new medicines. Experts estimate that routine use of organ-on-a-chip technology in drug research and development could bring about cost reductions of between ten and twenty-six per cent per new drug,[12] while using it to investigate toxicity could generate approximately $24 billion per year due to increased productivity.[13]

Of course, if effective treatments are found for our most common and debilitating diseases, there will be further savings. Dementia, for example, currently costs the UK economy £23.6 billion and this is predicted to rise to £59.4 billion by 2050. But if an intervention were found to delay the onset of dementia by just five years, this cost would reduce by thirty-six per cent.[14]

Human biology-based approaches are already contributing to the UK economy and this contribution is predicted to grow significantly over the next few years. A report by the Centre for Economics and Business Research found that in absolute terms, industry turnover in the sector grew by £452 million over the period 2017–19, reflecting an uptake in demand for the goods and services provided by human-relevant biotech industries. Labour productivity for firms operating in this sector also increased sharply over the same period. The report forecasts that industry employment will continue to grow and that by 2026, the sector will contribute £2.5 billion to UK GDP, an increase of seven hundred per cent from 2017.[15] The compound annual growth rate for the non-animal testing market is predicted to be higher than for the animal testing market over the next decade, and although the animal testing market is also expected to grow, experts expect this growth to be restrained. By contrast, the non-animal testing market is predicted to expand, offering lucrative opportunities for investors.[16,17,18]

Of course, if the UK government were to support the non-animal testing market, the potential for growth would be even greater.[14] This is something the majority of Britons would endorse, with several independent surveys showing solid support for a move towards non-animal methods in the UK. Seventy-five per cent of UK respondents feel that more should be done to find alternatives to animal experiments, according to the most recent government-commissioned survey, conducted in 2018.[19] Meanwhile, sixty per cent of British adults responding to a 2021 YouGov survey said they would back the government if it spent more on developing alternatives to using animals in medical research.[20] And a 2022 survey, this time conducted by market research company Savanta, revealed that seventy-nine per cent of British adults want the development and uptake of non-animal methods to be expedited, with seventy-six per cent wanting the UK to be a global leader in the transition to non-animal methods.[21]

For a period of almost twenty years, the UK government had commissioned a biennial survey to monitor public opinion about animal research, beginning in 1999. This survey had been showing fairly consistent levels of support for animal research until 2012, when acceptance levels began to decline. By 2018, it revealed that only forty-one per cent of the British public believed animal research to be important for human health.[19] Although the government has not recommissioned this survey since 2018, others have filled the vacuum. The

2021 YouGov survey revealed that over two-thirds (sixty-eight per cent) of British adults now want an end to animal experiments for medical research,[20] while the 2022 survey conducted by Savanta found that more than three-quarters (seventy-six per cent) of UK adults are 'very concerned' about the use of animals in scientific research and testing, with seventy-seven per cent wanting the UK government to phase out their use for these purposes.[21]

Petitions are another way of highlighting topics of public concern. In the UK, government e-petitions are frequently used to express people's disquiet; e-petitions reaching 10,000 signatures receive a parliamentary response, while those with at least 100,000 are considered for debate in parliament. Since October 2021, there have been no less than four e-petitions relating to the use of animals in science, all of which have attracted sufficient signatures to be debated in the UK parliament.[14]

And it's not just Britons who are concerned about the use of animals in research. US pollsters Gallup have recorded a steady decline in support for animal research among US citizens, from sixty-five per cent in 2001 to just fifty-two per cent in 2022,[22] while opposition to the practice is increasing. Meanwhile, in the EU, sixty-six per cent want all animal testing to end immediately.[23]

A transition from animal to human biology-based approaches would seem likely to result in public approval, then, in addition to economic and public health benefits. But there would be further advantages. Animal facilities are recognised to generate significant amounts of waste, such as animal excrement, bedding, excess feed and medical paraphernalia such as needles and syringes, as well as chemicals and the bodies of the animals themselves. Much of this is hazardous waste which tends to be incinerated, a process that generates harmful emissions, contributing to air pollution. Furthermore, animal facilities consume up to ten times more energy per square metre than offices, largely due to the need for ventilation and air conditioning.[24] And there are impacts on biodiversity too. The global biomedical and pharmaceutical research industry is having a major impact upon wild non-human primate populations; according to the International Union for the Conservation of Nature, the demand for long-tailed macaques in research is contributing to such a decline in their numbers that the species is now considered to be endangered.[25] All these impacts are concerning, but doubly so since they are the by-products of an unnecessary industry.

There are many benefits of moving away from the animal model paradigm, then. We know some of the reasons why greater progress has not been made in this direction, including lock-in mechanisms and dysfunctional regulations. But another reason for the lack of headway is that some scientists are fighting very hard to maintain the status quo. As the Belgian playwright Maurice Maeterlinck once observed, 'At every crossroads on the path that leads to the future, tradition has placed ten thousand men to guard the past.'

Ten thousand men

When UK lobbying group Understanding Animal Research realised that public support for animal experimentation was waning, their alarm bells started ringing. Believing that greater openness within the sector might win back public trust, the group quickly launched a consultation. According to the market research company conducting the consultation, participants felt that if the animal research sector wished to be regarded as open and transparent, it would need to subject itself to external scrutiny by those with an interest in animal welfare.[26] But despite participants having been assured that their views would influence the final report, the principle of public scrutiny failed to make its way into the 2014 'Concordat on openness in animal research'. Instead, its signatories – those UK institutions funding or conducting animal experiments – were simply asked to commit to engaging with the media and the public about their research.[27] As a result, they began to produce sleek web pages and highly sanitised films of laboratories, focusing on the supposed benefits of their research. As the Institute for Cancer Research states, 'the Concordat represents an opportunity for organisations like ours to demonstrate our commitment (to openness) publicly and to give our animal research an even higher profile when we communicate with the public'.[28] It is difficult, then, to see the Concordat as anything other than a public relations exercise aimed at improving the image of animal experimentation in order to win back public support and ensure its continuation.

Real openness would see members of the public being able to scrutinise and provide feedback on *proposed* research, not summaries of studies that have already been licensed, as is the case at present. In 2014, a government consultation took place with a view to amending Section 24 of the Animals (Scientific Procedures) Act 1986, which prevents the release of certain information such as the content of Project Licence applications.[29] The results of this consultation were never published, however, and no action has been taken.[30]

The precursor to the 'Concordat on openness in animal research' was the 2010 international Basel Declaration, which also calls for greater transparency. The Basel Declaration is explicit about its mission, insisting that 'necessary research involving animals, including non-human primates, be allowed now and in the future'.[31] The Basel Declaration Society was formed in 2011 to encourage adoption of the Declaration, morphing eventually into Animal Research Tomorrow which represents the interests of animal researchers all over the world. A primary goal is to 'engage in proactive communication about the continued importance of animal experimentation in the life sciences'.

'Animal research is here to stay,' stated Professor of Anatomy and Embryology, Rolf Zeller, opening Animal Research Tomorrow's 2020 conference. 'New developments like organoids are very important, great discoveries but cannot yet substitute animal experimentation.'[32]

Dr Kathrin Herrmann, Senior Associate at the Johns Hopkins Center for Alternatives to Animal Testing and Animal Protection Commissioner of Berlin, attended the conference. She was shocked.

'The researchers were basically discussing how can we get the next generation of scientists interested in animal experiments again. It looks like they are going in the other direction, they were really pushing back. There was one scientist who had founded an association to promote animal experimentation and defend it against those who oppose it.'

Animal Research Tomorrow makes much of the 3Rs,

as do similar organisations such as Understanding Animal Research. Yet one of the speakers at its 2020 conference, philosopher Matthias Eggel, argued that the benefit aspect of the harm–benefit analysis should refer solely to the creation of knowledge rather than societal benefits such as human health.[32] In practice, this would allow every single animal study to be approved by regulators, since all research produces knowledge of some kind, completely thwarting the principle of reduction. As support for animal research is declining, then, attempts are being made to move the goalposts.[33]

Initiatives aimed at bolstering support for animal experiments are springing up all over the world. Since 2016, every third Thursday in April has been Biomedical Research Awareness Day in the US, a day when 'the vital role animals play in the development of new treatments and cures for people and animals' is highlighted and 'honored', with posters, leaflets and other resources being supplied for use in schools and at community events. Meanwhile, as both the science and value of animal research are increasingly found wanting, animal users are shouting ever more loudly about how well regulated their research is.[34] A paper questioning the scientific validity of animal models that I wrote with Professor Merel Ritskes-Hoitinga,[35] for instance, prompted a response about ethics and the 3Rs,[36] but at no point did the authors address our actual argument, which was that animal models can never be scientifically valid. Questions about scientific validity are often met with declarations about the 3Rs: it's a case of deliberately missing the point, but the use of the 3Rs as a smokescreen does unfortunately succeed in putting some people on the back foot.

'The animal user industry is really fighting back,' says Kathrin Herrmann. 'They are seeing that the times are changing, and they have to explain themselves more and so they are very proactive in trying to continue the status quo. I was talking to a friend in the Netherlands, where the government is really strong on this exit plan to transition away from animal research, and she said the same is happening there, researchers are writing papers on how important animal experiments are.'

In fact, the Dutch government's plan to end the use of animals in safety testing met with so much opposition from the scientific community that its 2025 deadline was eventually abandoned.

'By letting go of the year, and with it resistance, progress can be made,' explained Dutch Minister Carola Schouten.

Since then, the EPA has also quietly dropped its 2035 target, stating that this had become the primary focus of discussion within the scientific community, as opposed to what actions could be taken. Consequently, the target date was removed.[37] What of the 2021 vote to establish an EU-wide action plan to phase out the use of animals in experiments? In June 2022, a meeting was held in the European Parliament to discuss this. It began with the chair, MEP Christian Ehler, stating that animal experiments were responsible for some of the most significant developments in human history and giving his view that the resolution was only being discussed because of changing social norms, completely ignoring the scientific imperative.

'I think that this institution still believes that for the given moment you can't replace animal testing,' he concluded after a long meeting. 'Not [unless] you want to seriously harm the medical interests of the European citizens but also science in general.'

A month later, the German research foundation Deutsche Forschungsgemeinschaft – the equivalent of the UK's Medical Research Council – stated that the EU's ambition to phase out animal experiments threatened scientific freedom and repeated the mantra that animal experiments will remain indispensable for the time being.[38] Increasingly, it feels as though both the brakes and accelerator are on at the same time. Three years after the Medical Research Council announced that research at its Mammalian Genetics Unit at Harwell was coming to an end, it invested £20 million into a venture at Harwell 'to integrate mouse models with human genetics'.[39]

Clearly, the transition to human-focused research is not going to be smooth. In their efforts to resist change, animal users are drawing upon a range of different arguments. One is that animals enable the study of a whole living organism. This is certainly true, but as we noted earlier, the generation of knowledge about how a new drug behaves in a rat or a monkey is not the same as providing information about how that drug will behave in humans. Another view increasingly put forward is that using larger animals will improve the likelihood of studies translating to humans.[40] For instance, a 2022 paper by members of another international lobbying group, Speaking of Research, argues that the use of non-human primates will lead to breakthroughs for Alzheimer's disease,[41] a condition that continues to lack effective treatments despite decades of research on rodents. This line of thought is worryingly simplistic, since the difficulties in extrapolating to humans do not just disappear because animals are bigger, or because they seem more like us on a superficial level; species differences will render this research unreliable whatever the size or appearance of the animal. Sometimes the argument seems to be as crude as 'we've always done it this way, so we'll carry on doing it this way'. In 2020, the Alzheimer's Association Business Consortium Think Tank convened a meeting of experts to explore the current state of animal models of Alzheimer's disease. Despite recognising that these models fail to adequately capture the complexity of human Alzheimer's disease, the bizarre conclusion was that they should continue to be developed.[42]

Arguments against using new approaches are also widespread. For example, some scientists protest that it's not possible to replace every single animal test with a non-animal equivalent.[43] But this has never been the aim.[44] It is neither necessary nor desirable to replace specific animal tests with like-for-like non-animal tests. Such a position only reinforces the misperception that animal tests are the gold standard, a mindset we need to discard.[45] A more logical – and creative – approach is to go back to the drawing board, focus on the research question and think about the best way of answering it.[46] Alternatively, animal use can often be completely bypassed simply by asking the research question in a different way.

A further, commonly cited objection is that the science behind human biology-based approaches is not yet ready. Kathy Archibald thinks this is nonsense.

'We've got human-relevant methodologies for pretty much everything,' she explains, 'but they're all sitting in tiny little labs without the money to scale up and develop and optimise. We may not have perfect, off-the-shelf tests for every single thing you can think of, but the fundamental science is there.'

Of course, innovative approaches must be fit for purpose, and research using them must be valid, reliable and accurately reported,[47] as with any research, but as Archibald states, the fundamental science is there. Moreover, data showing that these new approaches have a clear and unambiguous advantage over animal studies are accumulating.[48] Organs-on-a-chip are able to far outperform animal tests in detecting drugs that will be toxic to humans[13,49] and can replicate human physiological processes that animal models cannot, because the mechanisms are human-specific.[50] In other words, organ chips are doing things that animal models have never been, and will never be, capable of.

As John Maynard Keynes noted, 'The difficulty lies, not in the new ideas, but in escaping from the old ones.' Discarding outdated ideas is proving difficult, but we should not be surprised by the pushback. When the autonomy of scientists was threatened by the 1876 Cruelty to Animals Act, the first attempt anywhere in the world to regulate animal experimentation, scientists responded by forming lobbying groups, instigating campaigns, repeating slogans, closing ranks and raising objections. Animal researchers are doing the same now as they fight tooth and nail to resist progress. But as Professor Michael Balls points out, those who continue to exaggerate the value of animal experimentation and campaign to preserve the status quo will have to bear some responsibility for the harm this causes.[51]

A wicked problem

What can be done to make progress in the face of this pushback? In the field of social policy, problems that are highly complex and resistant to resolution have come to be known as 'wicked' problems. Wicked problems tend not to fall under the responsibility of any one organisation or government department[52] and they often involve multiple stakeholders with radically different world views and frames of reference. Recognising that the transition from animal to human-focused research is a wicked problem means acknowledging that it requires many levers to be pulled at the same time.

In the winter of 2018, researcher Paul Krijnen and Carine van Schie from the Dutch Burns Foundation asked for help. They wanted to investigate the efficacy of a drug for healing burn wounds but were curious about whether their research could be done without using mice. Experts in human biology-based research answered their call and what followed was an intense twenty-four-hour brainstorm, during which the scientists discussed the possibility of developing an in vitro model of burn wounds for testing the drug. This was the world's first 'Helpathon': a friendly collaboration dedicated to sharing expertise with the aim of answering research questions without using animals. Since then, a further six Dutch Helpathons have taken place, while the seventh – the UK's first – took place in late 2022. Participants report appreciating the creative, non-competitive atmosphere and gaining fresh perspectives and new insights.[53] Given that scientists from different paradigms often find it challenging to communicate, Helpathons provide an inspiring example of constructive dialogue.

Collaboration, then, is one lever that needs to be pulled. Another is education. The next generation can be encouraged to embrace human-based approaches through PhDs and other opportunities.[54] The charities Animal Free Research UK and FRAME (Fund for the Replacement of Animals in Medical Experiments) both fund annual summer schools for early career scientists in the UK, providing experience in animal-free laboratories and the opportunity to learn how new technologies may be employed to tackle human disease. Similar programmes are run elsewhere in Europe and in the US.[55]

Such initiatives, however, will only go so far because a transition on this scale requires top-down leadership as well as bottom-up initiatives. In other words, governments need to step up. The problem is, most politicians tend to regard animal research as a welfare issue only, understanding little about its scientific limitations or its impact on humans, and failing to grasp the transformative potential of new technologies. To make things worse, there is no obvious department within the UK government that might assume responsibility for a transition to human-focused science, partly explaining why progress here has stalled and why any advances are so slow and piecemeal. The UK All Party Parliamentary Group on Human Relevant Science intends to change this. Formed in 2020, it aims to bring MPs and peers together with scientists and other stakeholders to discuss and promote the uptake of human biology-based methodologies. Emphasising that the single most important requirement for a transition away from animal research is governmental support, the group recommends the creation of a dedicated ministerial position to spearhead the development of a road map and a properly resourced transition programme.[56,14]

A key element of such a programme would be an overhaul of the relevant regulations. As we have seen, the regulations governing animal research are utterly unfit for purpose and in need of complete reform. In place of the current box-ticking exercise, project licence applications need to be rigorously scrutinised to ensure that the only studies approved are those that are scientifically robust, have a strong likelihood of benefit, and cannot be conducted without using animals. To ensure robust science, statisticians will need to be involved in assessments. When considering the likelihood of benefit, the assumption should be that the research is *unlikely* to be beneficial, in line with the evidence. And, as a matter of urgency, Home Office inspectors and animal ethical review boards must acquire the expertise in human biology-based approaches necessary to properly assess whether a project can be conducted without using animals and to ensure compliance with Directive 2010/63/EU which insists upon the principle of replacement.[57] Members of the public should also have a much greater role in scrutinising proposed animal studies.[58,59]

UK regulations governing the safety testing of new drugs also need to be modernised, as they have been in the US. The UK regulator needs to urgently clarify that animal experiments are no longer the default option for preclinical safety testing and that non-animal, human-based approaches should, in accordance with Directive 2010/63/EU, be used by pharmaceutical companies wherever possible.[60]

The other essential component of a government-led

transition programme is to divert funding away from animal models and towards human-relevant research. No new funding is necessary; what already exists simply needs to be redistributed. Scientists follow the money, so for as long as this is poured into animal research, they will continue to conduct animal experiments. But if funding were diverted to human-focused approaches, they would rapidly find ways of conducting human-focused research. When, in 2013, it became illegal within the UK and EU to sell cosmetics that were newly tested on animals, or that contained ingredients tested on animals, scientists quickly began to develop the human-based tests that are now used instead. It is essential, therefore, that funding bodies use their resources to disincentivise animal studies and incentivise human-focused research.[56] For a start, funding for animal research in fields such as stroke, where decades of work have been entirely unproductive, should cease immediately. It's astonishing that national institutes, research councils and charities persist in financing such studies; indeed, it raises questions about whether such bodies actually audit the outcomes of the research they fund. Funding bodies and committees also need to gain expertise in human biology-based approaches. This would enable them to weed out proposals that could employ human-focused approaches instead of animal models, again ensuring compliance with Directive 2010/63/EU.

In short, a transition from animal to human-focused research needs to be government led, and it needs to focus on radically reforming funding and regulations. Only then will the potential of human biology-based approaches be realised.

The future

I stated at the beginning that this book would sidestep ethical issues and focus on the scientific limitations of animal studies. Yet, as we move deeper into a twenty-first century replete with superior technologies, we must recognise that the continued use of animals in preclinical research has become completely unethical. Unethical because, in the absence of benefits for humans, there remains no justification for using animals in research,[61,62] and because – as we have seen – animal research fails *us* spectacularly. Both aspects need to be acknowledged, with the ethical arguments expanding accordingly.[45,63,64] And of course it goes without saying that human biology-based approaches, because they avoid animal suffering and have more potential to relieve human anguish, are ethically superior to animal studies. Meanwhile, the ethical issues raised by human-based approaches, such as tissue donation and consent, appear relatively easy to resolve.

The animal research paradigm carries the weight of tradition, but it is now obsolete. As a research programme in decline, it cannot advance or successfully extend itself; instead, it rakes over its former successes and, far from breaking new ground, falls behind as it tries to resolve its key problem: lack of translation to humans. Attempts to revive or reinvent itself, as with the use of genetically modified animals, have failed to resolve this issue. Aware of its impending demise, claims about its indispensability are now frequently qualified by phrases such as 'for the time being', or 'for the foreseeable future'. The animal model paradigm is terminally ill, then, but for the time being is holding on, receiving inappropriate life support from funders, governments and scientists who appear unable to accept its fate. But there is no hope of recovery. There should be no more transfusions, no further sustenance. It's time to pull the plug.

ACKNOWLEDGEMENTS

My thanks go first and foremost to Safer Medicines Trust for funding the writing of this book and to all the Trust's supporters who made this possible. Special thanks for their remarkable and outstanding generosity go to Dr Christopher Anderegg, Beata and Andy Gajek and the far-sighted supporters who left bequests in their wills. Kathy Archibald, founder of Safer Medicines, believed in this book from its outset and has been my constant supporter and cheerleader. She not only came up with the title but read numerous versions of the manuscript and the book is all the better for her scrutiny. I am enormously grateful to my other readers too, including Dr Jan Turner, Dr Ricardo Blaug, Dr James Le Fanu, Professor Robert Matthews, Rebecca Ram and my husband Matt Penny, all of whom took the time to provide valuable feedback and much-needed encouragement. I am indebted to Dr Jonathan Balcombe and David Evans, both of whom gave me excellent advice early in the process and helped me shape the book proposal.

Rat Trap would not be the same without the wonderful contributions of those who have shared their time and expertise with me over the last couple of years. Enormous

ACKNOWLEDGEMENTS

thanks therefore go to Professor Merel Ritskes-Hoitinga, Dr Kathrin Herrmann, Professor Thomas Hartung, Dr Frances Cheng, Dr Katya Tsaioun, Professor Azra Raza, Dr Jarrod Bailey, Professor Lorna Harries, Professor Emeritus Robert Perlman, Professor Emeritus Geoff Pilkington and, once more, Dr Jan Turner and Kathy Archibald. Many of their contributions appear in the book, helping to bring it alive.

Matt has seen me through the highs and lows of writing and has been a continual supportive presence. Lily was my constant companion while I wrote, regularly persuading me to stretch my legs by taking her for a walk. Finally, thank you so much to Anne Elliott-Day, who proofread the text so carefully and conscientiously, and to Joshua, Sophie and Chelsea at Troubador. I am deeply grateful to you all.

END NOTES

Preface
1. The Raw. *Report – St Thomas's Hospital Medical School – London February 2015*. 28 Days Later Urban Exploration. 2015. https://www.28dayslater.co.uk/threads/st-thomass-hospital-medical-school-london-february-2015.94339/
2. Pound P, Ebrahim S, Sandercock P, Bracken MB, Roberts I. Where is the evidence that animal research benefits humans? *BMJ*. 2004;328(7438):514–517. doi:10.1136/bmj.328.7438.514
3. Responses to article: Where is the evidence that animal research benefits humans? *Br Med J*. 2004;328. https://www.bmj.com/content/328/7438/514/rapid-responses
4. Henderson M. Junk medicine: anti-vivisection campaigners. *The Times*. March 20, 2004. https://www.thetimes.co.uk/article/junk-medicine-anti-vivisection-campaigners-vjr6s7zq5n2
5. Highfield R. Experiments on animals should end, say doctors. *Telegraph*. February 27, 2004. https://www.telegraph.co.uk/news/uknews/1455415/Experiments-on-animals-should-end-say-doctors.html
6. Royal Society. *The Use of Non-Human Animals in Research: A Guide for Scientists*. 2004
7. BBC. Scientists doubt animal research. February 27, 2004. http://news.bbc.co.uk/1/hi/health/3489952.stm
8. Ritskes-Hoitinga M, Pound P. The role of systematic reviews in identifying the limitations of preclinical animal research,

2000–2022: Part 1. *J R Soc Med.* 2022;115(5):186–192. doi:10.1177/01410768221093551

9. Thomas D, Burns J, Audette J, Caroll A, Dow-Hygelund C, Hay M. *Clinical Development Success Rates 2006–2015.* 2016. https://www.bio.org/sites/default/files/legacy/bioorg/docs/Clinical Development Success Rates 2006–2015 – BIO, Biomedtracker, Amplion 2016.pdf

10. Alliance for Human Relevant Science. *Accelerating the Growth of Human Relevant Life Sciences in the United Kingdom.* A White Paper by the Alliance for Human Relevant Science. 2020. https//www.humanrelevantscience.org/wp-content/uploads/Accelerating-the-Growth-of-Human-Relevant-Sciences-in-the-UK_2020-final.pdf

11. Graham DJ, Campen D, Hui R, et al. Risk of acute myocardial infarction and sudden cardiac death in patients treated with cyclo-oxygenase 2 selective and non-selective non-steroidal anti-inflammatory drugs: nested case-control study. *Lancet.* 2005;365(9458):475–481. doi:10.1016/S0140-6736(05)17864-7

12. Kilkenny C, Parsons N, Kadyszewski E, et al. Survey of the quality of experimental design, statistical analysis and reporting of research using animals. *PLoS One.* 2009;4(11):e7824. doi:10.1371/journal.pone.0007824

13. Henderson VC, Demko N, Hakala A, et al. A meta-analysis of threats to valid clinical inference in preclinical research of sunitinib. *Elife.* 2015;4. doi:10.7554/eLife.08351

14. Hirst JA, Howick J, Aronson JK, et al. The need for randomization in animal trials: an overview of systematic reviews. *PLoS One.* 2014;9(6):e98856. doi:10.1371/journal.pone.0098856

15. Macleod MR, Lawson McLean A, Kyriakopoulou A, et al. Risk of bias in reports of in vivo research: a focus for improvement. *PLoS Biol.* 2015;13(10):e1002273. doi:10.1371/journal.pbio.1002273

16. Understanding Animal Research. *Concordat on Openness in Animal Research in the UK.* 2014. http://concordatopenness.

org.uk/wp-content/uploads/2017/04/Concordat-Final-Digital.pdf
17. Pound P, Bracken MB. Is animal research sufficiently evidence based to be a cornerstone of biomedical research? *BMJ*. 2014;348:g3387. doi:10.1136/bmj.g3387
18. Responses to article: Is animal research sufficiently evidence based to be a cornerstone of biomedical research? *BMJ*. 2014;348:g3387. https://www.bmj.com/content/348/bmj.g3387/rapid-responses
19. Ipsos Mori. *Public Attitudes to Animal Research in 2018*. 2019. https://www.ipsos.com/sites/default/files/ct/news/documents/2019-05/18-040753-01_ols_public_attitudes_to_animal_research_report_v3_191118_public.pdf
20. Blattner C. The recognition of animal sentience by the law. *J Anim Ethics*. 2019;9(2):121. doi:10.5406/janimalethics.9.2.0121
21. Pew Research Centre. *Most Americans Accept Genetic Engineering of Animals That Benefits Human Health, but Many Oppose Other Uses*. 2018. https://www.pewresearch.org/science/2018/08/16/most-americans-accept-genetic-engineering-of-animals-that-benefits-human-health-but-many-oppose-other-uses/
22. Animal Free Research. Poll: clear majority of Britons want end to animal testing in UK labs. 2021. https://www.animalfreeresearchuk.org/poll-clear-majority-of-britons-want-end-to-animal-testing-in-uk-labs/
23. Savanta ComRes. *Cruelty Free Europe – Animal Testing in the EU: A European Wide Survey among the Public to Gauge Perceptions of Animal Testing in the EU*. 2020. https://savanta.com/knowledge-centre/poll/cruelty-free-europe-animal-testing-in-the-eu/
24. Transitie proefdiervrije innovatie [Transition to animal free innovation]. *About TPI*. 2023. https://www.transitieproefdiervrijeinnovatie.nl/over-tpi
25. Haahr T. MEPs demand EU action plan to end the use of animals in research and testing. 2021. https://www.europarl.europa.eu/news/en/press-room/20210910IPR11926/meps-demand-eu-action-plan-to-end-the-use-of-animals-in-research-and-testing

26. Kuhn T. *The Structure of Scientific Revolutions*. University of Chicago Press; 1962

Part One: Trapped

CHAPTER 1: CAPTURE

1. Field T. *Experimental Animals: A Reality Fiction*. Solid Objects; 2016
2. Taylor K, Alvarez LR. An estimate of the number of animals used for scientific purposes worldwide in 2015. *Altern Lab Anim*. 2019;47(5–6):196–213. doi:10.1177/0261192919899853
3. Franco N. Animal experiments in biomedical research: a historical perspective. *Animals*. 2013;3(1):238–273. doi:10.3390/ani3010238
4. Conner A. Galen's analogy: animal experimentation and anatomy in the second century C.E. *Anthós*. 2017;8(1). doi:10.15760/anthos.2017.118
5. Allen Shotwell R. The revival of vivisection in the sixteenth century. *J Hist Biol*. 2013;46(2):171–197. doi:10.1007/s10739-012-9326-8
6. Burdon-Sanderson J, Klein E, Foster M, Brunton TB. *Handbook for the Physiological Laboratory*. Churchill; 1873
7. Richards S. Vicarious suffering, necessary pain: physiological method in late nineteenth-century Britain. In: Rupke N, ed. *Vivisection in Historical Perspective*. Croom Helm; 1987
8. Rupke N. *Vivisection in Historical Perspective*. Croom Helm; 1987
9. Elliot P. Vivisection and the emergence of experimental physiology in nineteenth-century France. In: Rupke N, ed. *Vivisection in Historical Perspective*. Croom Helm; 1987
10. Bernard C. *Introduction to the Study of Experimental Medicine*. 1865
11. Ramé ML. *The New Priesthood: A Protest against Vivisection*. 1893
12. Waddington I. The movement towards the professionalisation

of medicine. *Br Med J.* 1990;301(6754):688-690. doi:10.1136/bmj.301.6754.688
13. Bates A. *Anti-Vivisection and the Profession of Medicine in Britain.* Palgrave Macmillan; 2017
14. Rupke N. Pro-vivisection in England in the early 1880s: arguments and motives. In: Rupke N, ed. *Vivisection in Historical Perspective.* Croom Helm; 1987
15. Darwin C. *On the Origin of Species.* John Murray; 1859
16. Futuyma DJ. *Evolutionary Biology.* Sinauer Associates; 1986
17. Boddice R. Vivisecting Major: a Victorian gentleman scientist defends animal experimentation, 1876–1885. *Isis.* 2011;102:215–237
18. Sina I. *The Cannon of Medicine.* 1012. https://www.jameslindlibrary.org/ibn-sina-c-1012-ce-c-402-ah/
19. Anon. Practical observations and suggestions in medicine: second series. *Med Chir Rev.* 1847;5(9):151–166
20. Kirk R. The birth of the laboratory animal: biopolitics, animal experimentation and animal wellbeing. In: Chrulew M, Wadiwel DJ, eds. *Foucault and Animals.* Brill; 2017. doi:10.1163/9789004332232_002
21. Hamilton S. On the Cruelty to Animals Act, 15 August 1876. In: Felluga DF, ed. *BRANCH: Britain, Representation and Nineteenth-Century History.* 2013. http://www.branchcollective.org/?ps_articles=susan-hamilton-on-the-cruelty-to-animals-act-15-august-1876
22. Sharpey-Schafer E. History of the Physiological Society, 1876–1926. *J Physiol.* 1927;64(Suppl):1–76
23. Rudacille D. *The Scalpel and the Butterfly: The War between Animal Research and Animal Protection.* University of California Press; 2001
24. Wilks S. The Association for the Advancement of Medicine by Research. *Br Med J.* 1882. https://www.ncbi.nlm.nih.gov/pmc/articles/PMC2371713/pdf/brmedj04490-0035a.pdf
25. Owen PS. *Little Brown Dog.* Honno Press; 2021
26. Lind af Hageby E, Schartau LK. *The Shambles of Science: Extracts from the Diary of Two Students of Physiology.* E. Bell; 1903

27. LaFollette H, Shanks N. *Brute Science: Dilemmas of Animal Experimentation*. Routledge; 1997
28. Horton R. Jabs. *London Rev Books*. 1992;14(19)
29. Stahnisch FW. François Magendie (1783–1855). *J Neurol*. 2009;256(11):1950-1952. doi:10.1007/s00415-009-5291-3

CHAPTER 2: STUCK

1. Boddice R. Vivisecting Major: A Victorian gentleman scientist defends animal experimentation, 1876–1885. *Isis*. 2011;102:215–237
2. Tansey EM. Protection against dog distemper and dogs protection bills: the Medical Research Council and anti-vivisectionist protest, 1911–1933. *Med Hist*. 1994;38(1):1–26. doi:10.1017/S0025727300056027
3. MacDonald G. Mice worth their weight in gold: some extravagant pets. *The Harmsworth Monthly Pictorial Magazine*. 1898. https://www.afrma.org/pdfbooks/harmsworth1898s.pdf
4. Kirk RGW. 'Wanted-standard guinea pigs': standardisation and the experimental animal market in Britain ca. 1919–1947. *Stud Hist Philos Biol Biomed Sci*. 2008;39(3):280–291. doi:10.1016/j.shpsc.2008.06.002
5. Medical Research Council. *History of MRC*. 2022. https://www.ukri.org/about-us/mrc/who-we-are/our-history/
6. McNeill L. The history of breeding mice for science begins with a woman in a barn. *Smithson Mag*. 2018. https://www.smithsonianmag.com/science-nature/history-breeding-mice-science-leads-back-woman-barn-180968441/
7. Logan CA. Commercial rodents in America: standard animals, model animals, and biological diversity. *Brain Behav Evol*. 2019;93(2–3):70–81. doi:10.1159/000500073
8. Little C. US Science wars against an unknown enemy: cancer. *Life Mag*. March 1937:11–17
9. Logan CA. The legacy of Adolf Meyer's comparative approach: Worcester rats and the strange birth of the animal model. *Integr Physiol Behav Sci*. 2005;40(4):169–181. doi:10.4324/9781315135618-14

10. Logan C. The altered rationale for the choice of a standard animal in experimental psychology: Henry H. Donaldson, Adolf Meyer, and 'the' Albino Rat. *Hist Psychol.* 1999; 2(1):3–24
11. Preuss T. What animal models do to us (NIH Neuroscience Series Seminar). 2016. https://videocast.nih.gov/summary.asp?Live=17982&bhcp=1
12. LaFollette H, Shanks N. *Brute Science: Dilemmas of Animal Experimentation.* Routledge; 1997
13. Birke L, Arluke A, Michael M. *The Sacrifice. How Scientific Experiments Transform Animals and People.* Purdue University Press; 2007
14. Rowan AN, Loew FM. Animal research: a review of developments, 1950–2000. In: Salem DJ, Rowan AN, eds. *The State of the Animals 2001.* Humane Society Press; 2001:111–120. https://www.wellbeingintlstudiesrepository.org/cgi/viewcontent.cgi?article=1006&context=sota_2001
15. De Chadarevian S. Mice and the reactor: the 'genetics experiment' in 1950s Britain. *J Hist Biol.* 2006;39(4):707–735. doi:10.1007/s10739-006-9110-8
16. Schardein J. *Chemically Induced Birth Defects.* Marcel Dekker; 1985
17. Burnside R. *Safer Medicines.* 2006. ElstreeDV. https://safermedicines.org/safermedicines_video/
18. Frank J. Technological lock-in, positive institutional feedback, and research on laboratory animals. *Struct Chang Econ Dyn.* 2005;16(4):557-575. doi:10.1016/j.strueco.2004.11.001
19. Harris R. Drugs that work in mice often fail when tried in people. National Public Radio clip. 2017. https://www.npr.org/sections/health-shots/2017/04/10/522775456/drugs-that-work-in-mice-often-fail-when-tried-in-people?t=1603725158790&t=1603789668763
20. Latour B. Drawing things together. In: Lynch M, Woolgar S, eds. *Representations in Scientific Practice.* MIT Press; 1990
21. PETA. And the award goes to… Dr John Gluck! 2022. https://prime.peta.org/news/peta-2022-nanci-alexander-activist-award-dr-john-gluck/

22. Singer P. *Animal Liberation*. Harper Collins; 1975
23. Gluck JP. *Voracious Science and Vulnerable Animals*. University of Chicago Press; 2016
24. Elston MA. Attacking the foundations of modern medicine? Anti-vivisection protest and medical science. In: Keller D, Gabe J, Williams G, eds. *Challenging Medicine*. Routledge; 1994
25. Zhu F, Nair RR, Fisher EMC, Cunningham TJ. Humanising the mouse genome piece by piece. *Nat Commun*. 2019;10(1):1–13. doi:10.1038/s41467-019-09716-7
26. Home Office. *Annual Statistics of Scientific Procedures on Living Animals, Great Britain: 2021*. 2022. https://assets.publishing.service.gov.uk/government/uploads/system/uploads/attachment_data/file/1085383/annual-statistics-scientific-procedures-living-animals-2021_v8.pdf
27. The Business Research Company. *Animal Testing and Non-Animal Alternative Testing Market Players Invest in Novel Technologies as per The Business Research Company's Animal Testing and Non-Animal Alternative Testing Global Market Report 2022*. 2022. https://www.globenewswire.com/en/news-release/2022/01/26/2373712/0/en/Animal-Testing-And-Non-Animal-Alternative-Testing-Market-Players-Invest-In-Novel-Technologies-As-Per-The-Business-Research-Company-s-Animal-Testing-And-Non-Animal-Alternative-Testi.html
28. Bottini AA, Hartung T. Food for thought ... on the economics of animal testing. *ALTEX*. 2009;26(1):3–16. doi:10.14573/altex.2009.1.3
29. Macrotrends. *Charles River Laboratories Net Worth 2010–2022*. 2022. https://www.macrotrends.net/stocks/charts/CRL/charles-river-laboratories/net-worth
30. Charles River Laboratories. *Animal Models*. 2022. https://www.criver.com/site-search?s=wistar rat
31. The Jackson Laboratory. *Our Economic Impact*. 2022. https://www.jax.org/about-us/our-impact/economic-impact
32. Sturge G. *Animal Experiment Statistics*. House of Commons Library. 2022. https://commonslibrary.parliament.uk/research-briefings/sn02720/

33. Understanding Animal Research, Coalition for Medical Progress. *Medical Advances and Animal Research. The Contribution of Animal Science to the Medical Revolution: Some Case Histories.* 2007. https://www.understandinganimalresearch.org.uk/application/files/7016/4380/3819/medical-advances-and.pdf
34. McKeown T. *The Role of Medicine: Dream, Mirage or Nemesis?* Basil Blackwell; 1979

Part Two: In captivity

CHAPTER 3: BIAS AT THE BENCH

1. Sandercock P, Roberts I. Systematic reviews of animal experiments. *Lancet.* 2002;360(9333):586. doi:10.1016/S0140-6736(02)09812-4
2. Ritskes-Hoitinga M, Pound P. The role of systematic reviews in identifying the limitations of preclinical animal research, 2000–2022: Part 1. *J R Soc Med.* 2022;115(5):186–192. doi:10.1177/01410768221093551
3. Pound P, Ebrahim S, Sandercock P, Bracken MB, Roberts I. Where is the evidence that animal research benefits humans? *BMJ.* 2004;328(7438):514–517. doi:10.1136/bmj.328.7438.514
4. Chalmers I, Glasziou P. Avoidable waste in the production and reporting of research evidence. *Lancet.* 2009;374(9683):86–89. doi:10.1016/S0140-6736(09)60329-9
5. Ipsos Mori. *Public Attitudes to Animal Research in 2018.* 2019. https://www.ipsos.com/sites/default/files/ct/news/documents/2019-05/18-040753-01_ols_public_attitudes_to_animal_research_report_v3_191118_public.pdf
6. Reichlin TS, Vogt L, Würbel H. The researchers' view of scientific rigor – survey on the conduct and reporting of in vivo research. *PLoS One.* 2016;11(12):e0165999. doi:10.1371/journal.pone.0165999
7. Kilkenny C, Parsons N, Kadyszewski E, et al. Survey of the quality of experimental design, statistical analysis and reporting of research using animals. *PLoS One.* 2009;4(11):e7824. doi:10.1371/journal.pone.0007824

END NOTES

8. Henderson VC, Demko N, Hakala A, et al. A meta-analysis of threats to valid clinical inference in preclinical research of sunitinib. *Elife*. 2015;4. doi:10.7554/eLife.08351
9. Hirst JA, Howick J, Aronson JK, et al. The need for randomization in animal trials: an overview of systematic reviews. *PLoS One*. 2014;9(6):e98856. doi:10.1371/journal.pone.0098856
10. Macleod MR, Lawson McLean A, Kyriakopoulou A, et al. Risk of bias in reports of in vivo research: a focus for improvement. *PLoS Biol*. 2015;13(10):e1002273. doi:10.1371/journal.pbio.1002273
11. Fitzpatrick BG, Koustova E, Wang Y. Getting personal with the 'reproducibility crisis': interviews in the animal research community. *Lab Anim*. 2018;47:175–177
12. Scott S, Kranz JE, Cole J, et al. Design, power, and interpretation of studies in the standard murine model of ALS. *Amyotroph Lateral Scler*. 2008;9(1):4–15. doi:10.1080/17482960701856300
13. Chesler EJ, Wilson SG, Lariviere WR, Rodriguez-Zas SL, Mogil JS. Identification and ranking of genetic and laboratory environment factors influencing a behavioral trait, thermal nociception, via computational analysis of a large data archive. *Neurosci Biobehav Rev*. 2002;26(8):907–923. doi:10.1016/S0149-7634(02)00103-3
14. Jaric I, Voelkl B, Clerc M, et al. Rearing environment persistently modulates the phenotype of mice. *bioRxiv*. 2022. https://doi.org/10.1101/2022.02.11.480070
15. Everitt JI. The future of preclinical animal models in pharmaceutical discovery and development. *Toxicol Pathol*. 2015;43(1):70–77. doi:10.1177/0192623314555162
16. University of Bergen. Use of perfume in a lab animal facility. 2023. https://www.uib.no/en/rg/animalfacility/89275/use-perfume-lab-animal-facility
17. Crossley NA, Sena E, Goehler J, et al. Empirical evidence of bias in the design of experimental stroke studies. *Stroke*. 2008;39(3):929–934. doi:10.1161/STROKEAHA.107.498725

18. Schmidt-Pogoda A, Bonberg N, Koecke MHM, et al. Why most acute stroke studies are positive in animals but not in patients: a systematic comparison of preclinical, early phase, and Phase 3 clinical trials of neuroprotective agents. *Ann Neurol.* 2020;87(1):40–51. doi:10.1002/ana.25643
19. Lalu M, Leung GJ, Dong YY, et al. Mapping the preclinical to clinical evidence and development trajectory of the oncolytic virus talimogene laherparepvec (T-VEC): a systematic review. *BMJ Open.* 2019;9(12):e029475. doi:10.1136/bmjopen-2019-029475
20. Ioannidis JPA. Extrapolating from animals to humans. *Sci Transl Med.* 2012;4(151):151ps15. doi:10.1126/scitranslmed.3004631
21. Rosenblueth A, Wiener N. The role of models in science. *Philos Sci.* 1945;12(4):316–321
22. Balls M. It's time to reconsider the principles of humane experimental technique. *Altern Lab Anim.* 2020;48(1):40–46. doi:10.1177/0261192920911339
23. Seok J, Warren HS, Cuenca AG, et al. Genomic responses in mouse models poorly mimic human inflammatory diseases. *Proc Natl Acad Sci U S A.* 2013;110(9):3507–3512. doi:10.1073/pnas.1222878110
24. Warren HS. *Why Do We Use Mice to Study Human Diseases?* YouTube; 2013. https://www.youtube.com/watch?v=_V0LlyV1WWQ
25. Kolata G. Mice fall short as test subjects for some of humans' deadly ills. *New York Times.* February 11, 2013. https://www.nytimes.com/2013/02/12/science/testing-of-some-deadly-diseases-on-mice-mislead-report-says.html
26. Cassotta M, Pistollato F, Battino M. Rheumatoid arthritis research in the 21st century: limitations of traditional models, new technologies, and opportunities for a human biology-based approach. *ALTEX.* 2019;37(223–242). doi:10.14573/altex.1910011
27. Raza A. *The First Cell and the Human Costs of Pursuing Cancer to the Last.* Basic Books; 2019
28. Stanford SC. Some reasons why preclinical studies of psychiatric

disorders fail to translate: what can be rescued from the misunderstanding and misuse of animal 'models'? *Altern to Lab Anim.* 2020:026119292093987. doi:10.1177/0261192920939876

29. Genzel L, Adan R, Berns A, et al. How the COVID-19 pandemic highlights the necessity of animal research. *Curr Biol.* August 2020. doi:10.1016/j.cub.2020.08.030

30. Howells DW, Macleod MR. Evidence-based translational medicine. *Stroke.* 2013;44(5):1466–1471. doi:10.1161/STROKEAHA.113.000469

31. Vesterinen HM, Sena ES, ffrench-Constant C, Williams A, Chandran S, Macleod MR. Improving the translational hit of experimental treatments in multiple sclerosis. *Mult Scler J.* 2010;16(9):1044–1055. doi:10.1177/1352458510379612

32. Zeiss CJ, Allore HG, Beck AP. Established patterns of animal study design undermine translation of disease-modifying therapies for Parkinson's disease. *PLoS One.* 2017;12(2):e0171790. doi:10.1371/journal.pone.0171790

33. Zeeff SB, Kunne C, Bouma G, de Vries RB, te Velde AA. Actual usage and quality of experimental colitis models in preclinical efficacy testing. *Inflamm Bowel Dis.* 2016;22(6):1296–1305. doi:10.1097/MIB.0000000000000758

34. Malfait A-M, Little CB. On the predictive utility of animal models of osteoarthritis. *Arthritis Res Ther.* 2015;17(1):225. doi:10.1186/s13075-015-0747-6

35. Manuppello J, Sullivan K, Baker E. Acute toxicity 'six-pack' studies supporting approved new drug applications in the US, 2015–2018. *Regul Toxicol Pharmacol.* 2020;114:104666. doi:10.1016/j.yrtph.2020.104666

36. Leenaars C, Stafleu F, de Jong D, et al. A systematic review comparing experimental design of animal and human methotrexate efficacy studies for rheumatoid arthritis: lessons for the translational value of animal studies. *Animals.* 2020;10(6):1–21. doi:10.3390/ani10061047

37. Wiebers DO, Adams HP, Whisnant JP. Animal models of stroke: Are they relevant to human disease? *Stroke.* 1990;21(1):1–3. doi:10.1161/01.STR.21.1.1

38. Harris R. Drugs that work in mice often fail when tried in people. 2017. National Public Radio clip. https://www.npr.org/sections/health-shots/2017/04/10/522775456/drugs-that-work-in-mice-often-fail-when-tried-in-people?t=1603725158790&t=1603789668763
39. Pound P, Nicol CJ. Retrospective harm benefit analysis of pre-clinical animal research for six treatment interventions. *PLoS One.* 2018;13(3):e0193758. doi:10.1371/journal.pone.0193758
40. Holman C, Piper SK, Grittner U, et al. Where have all the rodents gone? The effects of attrition in experimental research on cancer and stroke. *PLoS Biol.* 2016;14(1):e1002331. doi:10.1371/journal.pbio.1002331
41. Head ML, Holman L, Lanfear R, Kahn AT, Jennions MD. The extent and consequences of P-hacking in science. *PLoS Biol.* 2015;13(3):e1002106. doi:10.1371/journal.pbio.1002106
42. Tsilidis KK, Panagiotou OA, Sena ES, et al. Evaluation of excess significance bias in animal studies of neurological diseases. *PLoS Biol.* 2013;11(7):e1001609. doi:10.1371/journal.pbio.1001609
43. Cohen D. Oxford TB vaccine study calls into question selective use of animal data. *BMJ.* January 10, 2018:j5845. doi:10.1136/bmj.j5845
44. Bruckner T, Wieschowski S, Heider M, et al. Measurement challenges and causes of incomplete results reporting of biomedical animal studies: results from an interview study. *PLoS One.* 2022;17(8):e0271976. doi:10.1371/journal.pone.0271976
45. Wieschowski S, Biernot S, Deutsch S, et al. Publication rates in animal research. Extent and characteristics of published and non-published animal studies followed up at two German university medical centres. *PLoS One.* 2019;14(11):1–8. doi:10.1371/journal.pone.0223758
46. van der Naald M, Wenker S, Doevendans PA, Wever KE, Chamuleau SAJ. Publication rate in preclinical research: a plea for preregistration. *BMJ Open Sci.* 2020;4(1):e100051. doi:10.1136/bmjos-2019-100051

47. Sena ES, van der Worp HB, Bath PMW, Howells DW, Macleod MR. Publication bias in reports of animal stroke studies leads to major overstatement of efficacy. *PLoS Biol.* 2010;8(3):e1000344. doi:10.1371/journal.pbio.1000344
48. Federico CA, Carlisle B, Kimmelman J, Fergusson DA. Late, never or non-existent: the inaccessibility of preclinical evidence for new drugs. *Br J Pharmacol.* 2014;171(18):4247–4254. doi:10.1111/bph.12771
49. Pound P. Animal models: problems and prospects. In: Fischer B, ed. *The Routledge Handbook of Animal Ethics.* Taylor & Francis; 2019
50. Cleary SJ, Pitchford SC, Amison RT, et al. Animal models of mechanisms of SARS-CoV-2 infection and COVID-19 pathology. *Br J Pharmacol.* 2020;(May):1–15. doi:10.1111/bph.15143
51. Pound P, Ram R. Are researchers moving away from animal models as a result of poor clinical translation in the field of stroke? An analysis of opinion papers. *BMJ Open Sci.* 2020;4(1):e100041. doi:10.1136/bmjos-2019-100041
52. Kringe L, Sena ES, Motschall E, et al. Quality and validity of large animal experiments in stroke: a systematic review. *J Cereb Blood Flow Metab.* June 23, 2020:0271678X2093106. doi:10.1177/0271678X20931062
53. Herrmann K. *Expert Panel Discussion on the Future of Research and Testing in the European Union & Beyond.* August 25, 2022. https://www.youtube.com/watch?v=-eqrt28xyrc
54. Kousholt BS, Præstegaard KF, Stone JC, et al. Reporting quality in preclinical animal experimental research in 2009 and 2018: a nationwide systematic investigation. *PLoS One.* 2022;17(11):e0275962. doi:10.1371/journal.pone.0275962
55. Fisher M, Feuerstein G, Howells DW, et al. Update of the Stroke Therapy Academic Industry Roundtable Preclinical Recommendations. *Stroke.* 2009;40(6):22442250. doi:10.1161/STROKEAHA.108.541128
56. Savitz SI, Fisher M. Future of neuroprotection for acute

stroke: in the aftermath of the SAINT trials. *Ann Neurol.* 2007;61(5):396–402. doi:10.1002/ana.21127

57. Pound P, Ritskes-Hoitinga M. Is it possible to overcome issues of external validity in preclinical animal research? Why most animal models are bound to fail. *J Transl Med.* 2018;16(1):304. doi:10.1186/s12967-018-1678-1

CHAPTER 4: ELEPHANT IN THE LAB

1. Pound P, Ram R. Are researchers moving away from animal models as a result of poor clinical translation in the field of stroke? An analysis of opinion papers. *BMJ Open Sci.* 2020;4(1):e100041. doi:10.1136/bmjos-2019-100041
2. Understanding Animal Research. *FAQs.* 2022. https://www.understandinganimalresearch.org.uk/what-is-animal-research/faqs/
3. Yokogi K, Goto Y, Otsuka M, et al. Neuromedin U-deficient rats do not lose body weight or food intake. *Sci Rep.* 2022;12(1):17472. doi:10.1038/s41598-022-21764-6
4. Coleman RA. Drug discovery and development tomorrow – changing the mindset. *ATLA Altern to Lab Anim.* 2009;37(Suppl. 1):1–4. doi:10.1177/026119290903701s02
5. LaFollette H, Shanks N. *Brute Science: Dilemmas of Animal Experimentation.* Routledge; 1997.
6. Futuyma DJ. *Evolutionary Biology.* Sinauer Associates; 1986
7. Wall RJ, Shani M. Are animal models as good as we think? *Theriogenology.* 2008;69(1):2–9. doi:10.1016/j.theriogenology.2007.09.030
8. Bailey J, Taylor K. Non-human primates in neuroscience research: the case against its scientific necessity. *ATLA Altern to Lab Anim.* 2016;44(1):43–69. doi:10.1177/026119291604400101
9. Balls M. It's time to reconsider the principles of humane experimental technique. *Altern Lab Anim.* 2020;48(1):40–46. doi:10.1177/0261192920911339
10. Perlman RL. Mouse models of human disease: an evolutionary perspective. *Evol Med Public Heal.* April 27, 2016:eow014. doi:10.1093/emph/eow014

11. Cairns-Smith AG. *Seven Clues to the Origin of Life*. Cambridge University Press; 1985
12. Logan CA. The legacy of Adolf Meyer's comparative approach: Worcester rats and the strange birth of the animal model. *Integr Physiol Behav Sci*. 2005;40(4):169–181. doi:10.4324/9781315135618-14
13. Preuss T. What animal models do to us (NIH Neuroscience Series Seminar). 2016. https://videocast.nih.gov/summary.asp?Live=17982&bhcp=1
14. Veening-Griffioen D, Ferreira G, Boon W, et al. Tradition, not science, is the basis of animal model selection in translational and applied research. *ALTEX*. 2020;37:1–14. doi:10.14573/altex.2003301
15. Akhtar A. The flaws and human harms of animal experimentation. *Cambridge Q Healthc Ethics*. 2015;24(4):407–419. doi:10.1017/S0963180115000079
16. Geerts H. Of mice and men. Bridging the translational disconnect in CNS drug discovery. *CNS Drugs*. 2009;23(1):915–926.
17. Langley GR. Considering a new paradigm for Alzheimer's disease research. *Drug Discov Today*. 2014;19(8):1114–1124. doi:10.1016/j.drudis.2014.03.013
18. LaFollette H, Shanks N. Animal experimentation: the legacy of Claude Bernard. *Int Stud Philos Sci*. 1994;8(3):195–210. doi:10.1080/02698599408573495
19. Sina I. *The Cannon of Medicine*. 1012. https://www.jameslindlibrary.org/ibn-sina-c-1012-ce-c-402-ah/

CHAPTER 5: VITAL AND INDISPENSABLE?

1. Home Office. *Annual Statistics of Scientific Procedures on Living Animals, Great Britain: 2021*. 2022. https://assets.publishing.service.gov.uk/government/uploads/system/uploads/attachment_data/file/1085383/annual-statistics-scientific-procedures-living-animals-2021_v8.pdf
2. The Royal Society. *The Use of Animals in Research*. https://royalsociety.org/news/2012/use-animals-research/
3. Matthews RAJ. Medical progress depends on animal models

– doesn't it? *J R Soc Med.* 2008;101(2):95–98. doi:10.1258/jrsm.2007.070164
4. House of Lords. *Select Committee on Animals in Scientific Procedures.* 2002. https://publications.parliament.uk/pa/ld200102/ldselect/ldanimal/150/15001.htm
5. Page K. Cambridge primate research centre comes under scrutiny. *Lancet Neurol.* 2003;2(3):136. doi:10.1016/S1474-4422(03)00335-1
6. Understanding Animal Research. *Forty Reasons Why We Need Animals in Research.* 2021. https://www.understandinganimalresearch.org.uk/why/forty-reasons-why-we-need-animals-in-research/
7. Akhtar A. Covid-19 research: with or without animals? (Online webinar). 2020. https://www.youtube.com/watch?v=HIpUfNhMM40
8. Comroe JH, Dripps RD. Scientific basis for the support of biomedical science. *Science.* 1976;192(4235):105–111. doi:10.1126/science.769161
9. Understanding Animal Research, Coalition for Medical Progress. *Medical Advances and Animal Research. The Contribution of Animal Science to the Medical Revolution: Some Case Histories.* 2007. https://www.understandinganimalresearch.org.uk/application/files/7016/4380/3819/medical-advances-and.pdf
10. Greek R, Swingle Greek J. *Sacred Cows and Golden Geese.* Continuum; 2000
11. McKeown T. *The Role of Medicine: Dream, Mirage or Nemesis?* Basil Blackwell; 1979
12. Burnet M. *Genes, Dreams and Realities.* Penguin Books; 1971
13. Coleman RA. Drug discovery and development tomorrow – changing the mindset. *ATLA Altern to Lab Anim.* 2009;37(Suppl. 1):1–4. doi:10.1177/026119290903701s02
14. LaFollette H, Shanks N. *Brute Science: Dilemmas of Animal Experimentation.* Routledge; 1997.
15. Walport M. *Animal Research: Then and Now.* 80th Stephen Paget Memorial Lecture (Transcript of speech). 2016. https://www.understandinganimalresearch.org.uk/resources/video-

library/80th-paget-lecture-2016/
16. Understanding Animal Research. *For Human Health*. 2021. https://www.understandinganimalresearch.org.uk/why/human-health/
17. Reines BP. On the locus of medical discovery. *J Med Philos*. 1991;16(2):183–209. doi:10.1093/jmp/16.2.183
18. Cade J. Lithium salts in the treatment of psychotic excitement. *Med J Aust*. 1949;36:349–352.
19. Martin P, Brown N, Kraft A. From bedside to bench? Communities of promise, translational research and the making of blood stem cells. *Sci Cult (Lond)*. 2008;17(1):29–41. doi:10.1080/09505430701872921
20. Horn J, de Haan RJ, Vermeulen M, Luiten PGM, Limburg M. Nimodipine in animal model experiments of focal cerebral ischemia. *Stroke*. 2001;32(10):2433–2438. doi:10.1161/hs1001.096009
21. Pound P, Nicol CJ. Retrospective harm benefit analysis of pre-clinical animal research for six treatment interventions. *PLoS One*. 2018;13(3):e0193758. doi:10.1371/journal.pone.0193758
22. Choi S, O'Connell L, Min K, et al. Efficacy of vardenafil and sildenafil in facilitating penile erection in an animal model. *J Androl*. 2013. doi:https://doi.org/10.1002/j.1939-4640.2002.tb02239.x
23. Perel P, Roberts I, Sena E, et al. Comparison of treatment effects between animal experiments and clinical trials: systematic review. *BMJ*. 2007;334(7586):197. doi:10.1136/bmj.39048.407928.BE
24. Leenaars CHC, Kouwenaar C, Stafleu FR, et al. Animal to human translation: a systematic scoping review of reported concordance rates. *J Transl Med*. 2019;17(1):223. doi:10.1186/s12967-019-1976-2
25. Grant J. Evaluating 'payback' on biomedical research from papers cited in clinical guidelines: applied bibliometric study. *BMJ*. 2000;320(7242):1107–1111. doi:10.1136/bmj.320.7242.1107

26. Carvalho C, Peste F, Marques TA, Knight A, Vicente LM. The contribution of rat studies to the current knowledge of major depressive disorder: results from citation analysis. *Front Psychol.* 2020;11:1–8. doi:10.3389/fpsyg.2020.01486
27. Carvalho C, Varela SAM, Bastos LF, et al. The relevance of in silico, in vitro and non-human primate based approaches to clinical research on major depressive disorder. *Altern to Lab Anim.* 2019;47(3–4):128–139. doi:10.1177/0261192919885578
28. Knight A. The poor contribution of chimpanzee experiments to biomedical progress. *J Appl Anim Welf Sci.* 2007;10(4):281-308. doi:10.1080/10888700701555501
29. Öztürk A, Ersan Ö. Are the lives of animals well-spent in laboratory science research? A study of orthopaedic animal studies in Turkey. *Clin Orthop Relat Res.* 2020;478(9):1965–1970. doi:10.1097/CORR.0000000000001335
30. Ruan Y, Robinson NB, Khan FM, et al. The translation of surgical animal models to human clinical research: a cross sectional study. *Int J Surg.* 2020;77:25–29. doi:10.1016/j.ijsu.2020.03.023
31. Leopold SS. Editor's spotlight/take 5: Are the lives of animals well-spent in laboratory science research? A study of orthopaedic animal studies in Turkey. *Clin Orthop Relat Res.* 2020;478(9):1961–1964. doi:10.1097/CORR.0000000000001420
32. Gluck JP. *Voracious Science and Vulnerable Animals.* University of Chicago Press; 2016
33. Lockwood A. *Test Subjects.* 2019. https://testsubjectsfilm.com/

CHAPTER 6: DESPERATE PATIENTS

1. Pound P, Gompertz P, Ebrahim S. A patient-centred study of the consequences of stroke. *Clin Rehabil.* 1998;12(3):255–264. doi:10.1191/026921598666856867
2. Pound P, Gompertz P, Ebrahim S. Illness in the context of older age: the case of stroke. *Sociol Heal Illn.* 1998;20(4):489–506. doi:10.1111/1467-9566.00112
3. Pound P, Bury M, Gompertz P, Ebrahim S. Views of survivors

of stroke on benefits of physiotherapy. *Qual Health Care.* 1994;3(2):69–74. doi:10.1136/qshc.3.2.69
4. Pound P, Bury M, Gompertz P, Ebrahim S. Stroke patients' views on their admission to hospital. *BMJ.* 1995;311(6996):18–22. doi:10.1136/bmj.311.6996.18
5. Pound P, Ram R. Are researchers moving away from animal models as a result of poor clinical translation in the field of stroke? An analysis of opinion papers. *BMJ Open Sci.* 2020;4(1):e100041. doi:10.1136/bmjos-2019-100041
6. Intercollegiate Stroke Working Party. *SSNAP (Sentinel Stroke National Audit Programme) Annual Public Report 2018.* 2019. https://www.strokeaudit.org/Documents/National/Clinical/Apr2017Mar2018/Apr2017Mar2018-AnnualReport.aspx
7. Howells DW, Sena ES, Macleod MR. Bringing rigour to translational medicine. *Nat Rev Neurol.* 2014;10(1):37
8. O'Collins VE, Macleod MR, Donnan GA, Horky LL, van der Worp BH, Howells DW. 1,026 experimental treatments in acute stroke. *Ann Neurol.* 2006;59 (3):467–477. doi:10.1002/ana.20741
9. Howells DW, Sena ES, O'Collins V, Macleod MR. Improving the efficiency of the development of drugs for stroke. *Int J Stroke.* 2012;7(5):371–377. doi:10.1111/j.1747-4949.2012.00805.x
10. Cummings JL, Morstorf T, Zhong K. Alzheimer's disease drug-development pipeline: few candidates, frequent failures. *Alzheimers Res Ther.* 2014;6(4):37. doi:10.1186/alzrt269
11. Alteri E, Guizzaro L. Be open about drug failures to speed up research. *Nature.* 2018;563(7731):317–319. doi:10.1038/d41586-018-07352-7
12. Triunfol M, Gouveia FC. What's not in the news headlines or titles of Alzheimer disease articles? #InMice. *PLoS Biol.* 2021;19(6):1–15. doi:10.1371/journal.pbio.3001260
13. Pistollato F, Bernasconi C, McCarthy J, et al. Alzheimer's disease, and breast and prostate cancer research: translational failures and the importance to monitor outputs and impact of funded research. *Animals.* 2020;10(7):1194. doi:10.3390/ani10071194

14. Thomas D, Burns J, Audette J, Caroll A, Dow-Hygelund C, Hay M. *Clinical Development Success Rates 2006–2015.* 2016. https://www.bio.org/sites/default/files/legacy/bioorg/docs/Clinical Development Success Rates 2006–2015 – BIO, Biomedtracker, Amplion 2016.pdf
15. Hwang TJ, Carpenter D, Lauffenburger JC, Wang B, Franklin JM, Kesselheim AS. Failure of investigational drugs in late stage clinical development and publication of trial results. *JAMA Intern Med.* 2016;176(12):1826. doi:10.1001/jamainternmed.2016.6008
16. Harrison RK. Phase II and Phase III failures: 2013–2015. *Nat Rev Drug Discov.* 2016;15(12):817–818. doi:10.1038/nrd.2016.184
17. Mukherjee S. *The Emperor of All Maladies: A Biography of Cancer.* Fourth Estate; 2010
18. Raza A. *Cancer Research: A New Paradigm.* Lecture given to Royal Society of Medicine October 27, 2022. https://www.youtube.com/watch?v=ViRa1W3z0-w
19. Raza A. *The First Cell and the Human Costs of Pursuing Cancer to the Last.* Basic Books; 2019
20. Raza A. *21st Century Innovations in Alternatives to Animals in Biomedical Research: Congressional Panel on the Humane Research & Testing Act.* 2021. https://mailchi.mp/cc299895613b/shh1wdpt7w-538419#congress
21. Pound P. Animal models and the search for drug treatments for traumatic brain injury. In: Johnson LSM, Fenton A, Shriver A, eds. *Neuroethics and Nonhuman Animals.* Springer Nature; 2020
22. Malfait A-M, Little CB. On the predictive utility of animal models of osteoarthritis. *Arthritis Res Ther.* 2015;17(1):225. doi:10.1186/s13075-015-0747-6
23. Vesterinen HM, Sena ES, ffrench-Constant C, Williams A, Chandran S, Macleod MR. Improving the translational hit of experimental treatments in multiple sclerosis. *Mult Scler J.* 2010;16(9):1044–1055. doi:10.1177/1352458510379612
24. Parker JL, Kohler JC. The success rate of new drug development in clinical trials: Crohn's disease. *J Pharm Pharm*

Sci. 2010;13(2):191–197. doi:10.18433/j39014
25. Perrin S. Preclinical research: make mouse studies work. *Nature.* 2014;507:423–425
26. McNamee K, Williams R, Seed M. Animal models of rheumatoid arthritis: How informative are they? *Eur J Pharmacol.* 2015;759:278–286. doi:10.1016/j.ejphar.2015.03.047
27. Mullane K, Williams M. Animal models of asthma: Reprise or reboot? *Biochem Pharmacol.* 2014;87(1):131–139. doi:10.1016/j.bcp.2013.06.026
28. Sheets RL, Zhou TQ, Knezevic I. Review of efficacy trials of HIV-1/AIDS vaccines and regulatory lessons learned. A review from a regulatory perspective. *Biologicals.* 2016;44(2):73–89. doi:10.1016/j.biologicals.2015.10.004
29. Czéh B, Fuchs E, Wiborg O, Simon M. Animal models of major depression and their clinical implications. *Prog Neuro-Psychopharmacology Biol Psychiatry.* 2016;64:293–310. doi:10.1016/j.pnpbp.2015.04.004
30. Potashkin JA, Blume SR, Runkle NK. Limitations of animal models of Parkinson's disease. *Parkinsons Dis.* 2011; December 20. doi:10.4061/2011/658083
31. Seok J, Warren HS, Cuenca AG, et al. Genomic responses in mouse models poorly mimic human inflammatory diseases. *Proc Natl Acad Sci U S A.* 2013;110(9):3507_3512. doi:10.1073/pnas.1222878110
32. Roep BO, Atkinson M, Von Herrath M. Satisfaction (not) guaranteed: re-evaluating the use of animal models of type 1 diabetes. *Nat Rev Immunol.* 2004;4(12):989_997. doi:10.1038/nri1502
33. Animal Research Tomorrow. *Basel Declaration.* 2022. https://animalresearchtomorrow.org/en/basel-declaration
34. McKeown T. *The Role of Medicine: Dream, Mirage or Nemesis?* Basil Blackwell; 1979
35. Bynum B. The McKeown thesis. *Lancet.* 2008;371(9613):644–645. doi:10.1016/S0140-6736(08)60292-5
36. Colgrove J. The McKeown thesis, a historical controversy. *Heal Policy Ethics.* 2002;92(5):725–729

37. LaFollette H, Shanks N. *Brute Science: Dilemmas of Animal Experimentation*. Routledge; 1997
38. Bailey J, Balls M. Clinical impact of high-profile animal-based research reported in the UK national press. *BMJ Open Sci*. 2020;4(1):e100039. doi:10.1136/bmjos-2019-100039
39. Contopoulos-Ioannidis DG, Ntzani EE, Ioannidis JPA. Translation of highly promising basic science research into clinical applications. *Am J Med*. 2003;114(6):477–484. doi:10.1016/S0002-9343(03)00013-5
40. Hackam D, Redelmeier D. Translation of research evidence from animals to humans. *JAMA*. 2006;296(14):1731–1732
41. Marshall LJ, Triunfol M, Seidle T. Patient-derived xenograft vs. organoids: a preliminary analysis of cancer research output, funding and human health impact in 2014–2019. *Animals*. 2020;10(1923). doi:10.3390/ani10101923
42. Morgan M, Barry CA, Donovan JL, Sandall J, Wolfe CDA, Boaz A. Implementing 'translational' biomedical research: convergence and divergence among clinical and basic scientists. *Soc Sci Med*. 2011;73(7):945–952. doi:10.1016/j.socscimed.2011.06.060
43. Wooding S, Hanney S, Pollitt A, Buxton M, Grant J. Project Retrosight: understanding the returns from cardiovascular and stroke research: the policy report. *Rand Heal Q*. 2011;1(1):16
44. Wooding S, Pollitt A, Castle-Clarke S, et al. Mental Health Retrosight: understanding the returns from research (lessons from schizophrenia): policy report. *Rand Heal Q*. 2014;4(1)
45. Guthrie S, Kirtley A, Garrod B, Pollitt A, Grant J, Wooding S. *A 'DECISIVE' Approach to Research Funding Lessons from Three Retrosight Studies*. 2016. https://www.rand.org/pubs/research_reports/RR1132.html
46. Rothwell PM. Funding for practice-oriented clinical research. *Lancet*. 2006;368(9532):262–266. doi:10.1016/S0140-6736(06)69010-7
47. Dirnagl U. Thomas Willis Lecture. *Stroke*. 2016;47(8):2148–2153. doi:10.1161/STROKEAHA.116.013244
48. Gorelick PB. The global burden of stroke: persistent and

disabling. *Lancet Neurol.* 2019;18(5):417–418. doi:10.1016/S1474-4422(19)30030-4
49. Stroke Unit Trialists' Collaboration. Organised inpatient (stroke unit) care for stroke. *Cochrane Database Syst Rev.* September 11, 2013. doi:10.1002/14651858.CD000197.pub3
50. Kaste M. Use of animal models has not contributed to development of acute stroke therapies. *Stroke.* 2005;36(10):2323–2324. doi:10.1161/01.STR.0000179037.82647.48
51. Keating P, Cambrosio A. The new genetics and cancer: the contributions of clinical medicine in the era of biomedicine. *J Hist Med Allied Sci.* 2001;56(4):321–352
52. Godlee F. How predictive and productive is animal research? *BMJ.* 2014;348:g3719. doi:10.1136/bmj.g3719

CHAPTER 7: THE REAL GUINEA PIGS

1. Knight K. The lifelong shadow hanging over the Elephant Man drug trial victims after the human guinea pigs were left horribly disfigured and fighting for their lives. *Mail Online.* February 17, 2017. https://www.dailymail.co.uk/news/article-4236132/Lifelong-shadow-hanging-Elephant-Man-drug-trial-men.html.
2. Bradford D. I ran a medical trial that went wrong. *Guardian.* April 22, 2016. https://www.theguardian.com/lifeandstyle/2016/apr/22/experience-i-ran-medical-trial-that-went-wrong.
3. Suntharalingam G, Perry MR, Ward S, et al. Cytokine storm in a Phase 1 trial of the anti-CD28 monoclonal antibody TGN1412. *N Engl J Med.* 2006;355(10):1018–1028. doi:10.1056/NEJMoa063842
4. Archibald K, Coleman R, Drake T. Replacing animal tests to improve safety for humans. In: Herrmann K, Jayne K, eds. *Animal Experimentation: Working Towards a Paradigm Change.* Brill; 2019:417–442. doi:10.1163/9789004391192_019
5. Attarwala H. TGN1412: from discovery to disaster. *J Young Pharm.* 2010;2(3):332–336. doi:10.4103/0975-1483.66810
6. National Institute for Biological Standards and Control.

TGN1412 – learning from a clinical trials disaster. 2020. https://www.nibsc.org/about_us/worldwide_impact/tgn1412.aspx
7. Kerbrat A, Ferré J-C, Fillatre P, et al. Acute neurologic disorder from an inhibitor of fatty acid amide hydrolase. *N Engl J Med.* 2016;375(18):1717–1725. doi:10.1056/NEJMoa1604221
8. van Esbroeck ACM, Janssen APA, Cognetta AB, et al. Activity-based protein profiling reveals off-target proteins of the FAAH inhibitor BIA 10-2474. *Science.* 2017;356(6342):1084–1087. doi:10.1126/science.aaf7497
9. US Food and Drug Administration. *Guidance for Industry Estimating the Maximum Safe Starting Dose in Initial Clinical Trials for Therapeutics in Adult Healthy Volunteers.* 2005. https://www.fda.gov/regulatory-information/search-fda-guidance-documents/estimating-maximum-safe-starting-dose-initial-clinical-trials-therapeutics-adult-healthy-volunteers
10. Manning F, Swartz M. (eds.) *Institute of Medicine (US) Committee to Review the Fialuridine (FIAU/FIAC) Clinical Trials.* National Academies Press (US); 1995
11. Bagot M. Trial and error: thousands left seriously ill or disabled by clinical tests of new drugs. *Daily Mirror.* June 11, 2015. https://www.mirror.co.uk/news/uk-news/trial-error-thousands-left-seriously-5867052
12. Ioannidis JPA. Adverse events in randomized trials. *Arch Intern Med.* 2009;169(19):1737. doi:10.1001/archinternmed.2009.313
13. Saxena R, Wijnhoud AD, Carton H, et al. Controlled safety study of a hemoglobin-based oxygen carrier, DCLHb, in acute ischemic stroke. *Stroke.* 1999;30(5):993–996. doi:10.1161/01.STR.30.5.993
14. Enlimomab Acute Stroke Trial Investigators. Use of anti-ICAM-1 therapy in ischemic stroke: results of the Enlimomab Acute Stroke Trial. *Neurology.* 2001;57(8):1428–1434. doi:10.1212/WNL.57.8.1428
15. Davis SM, Lees KR, Albers GW, et al. Selfotel in acute ischemic stroke: possible neurotoxic effects of an NMDA antagonist.

Stroke. 2000;31(2):347–354. doi:10.1161/01.STR.31.2.347

16. Tirilazad International Steering Committee. Tirilazad for acute ischaemic stroke. *Cochrane Database Syst Rev*. 2001;4.

17. Hwang TJ, Lauffenburger JC, Franklin JM, Kesselheim AS. Temporal trends and factors associated with cardiovascular drug development, 1990 to 2012. *JACC Basic to Transl Sci*. 2016;1(5):301–308. doi:10.1016/j.jacbts.2016.03.012

18. Barter PJ, Caulfield M, Eriksson M, et al. Effects of torcetrapib in patients at high risk for coronary events. *N Engl J Med*. 2007;357(21):2109–2122. doi:10.1056/NEJMoa0706628

19. Hermann M, Ruschitzka FT. The hypertension peril: lessons from CETP inhibitors. *Curr Hypertens Rep*. 2009;11(1):76–80. doi:10.1007/s11906-009-0014-9

20. Pollack A. Pfizer to lay off 10,000 workers. *New York Times*. January 22, 2007. https://www.nytimes.com/2007/01/22/business/22cnd-pfizer.html.

21. Editorial. Fourth child dies in gene therapy trial. *Pharma Manuf*. September 2021. https://www.pharmamanufacturing.com/development/clinical-trials/news/11291158/fourth-child-dies-in-gene-therapy-trial

22. Fidler B. Third patient dies in halted study of Audentes gene therapy. *Biopharmadive*. 2020. https://www.biopharmadive.com/news/audentes-gene-therapy-patient-deaths/580670/

23. Childers MK, Joubert R, Poulard K, et al. Gene therapy prolongs survival and restores function in murine and canine models of myotubular myopathy. *Sci Transl Med*. 2014;6(220). doi:10.1126/scitranslmed.3007523

24. Hwang TJ, Carpenter D, Lauffenburger JC, Wang B, Franklin JM, Kesselheim AS. Failure of investigational drugs in late stage clinical development and publication of trial results. *JAMA Intern Med*. 2016;176(12):1826. doi:10.1001/jamainternmed.2016.6008

25. Harrison RK. Phase II and Phase III failures: 2013–2015. *Nat Rev Drug Discov*. 2016;15(12):817–818. doi:10.1038/nrd.2016.184

26. Siramshetty VB, Nickel J, Omieczynski C, Gohlke BO,

Drwal MN, Preissner R. WITHDRAWN – a resource for withdrawn and discontinued drugs. *Nucleic Acids Res.* 2016;44(D1):D1080–D1086. doi:10.1093/nar/gkv1192

27. Graham DJ, Campen D, Hui R, et al. Risk of acute myocardial infarction and sudden cardiac death in patients treated with cyclo-oxygenase 2 selective and non-selective non-steroidal anti-inflammatory drugs: nested case-control study. *Lancet.* 2005;365(9458):475–481. doi:10.1016/S0140-6736(05)17864-7

28. Levesque LE. Time variations in the risk of myocardial infarction among elderly users of COX-2 inhibitors. *Can Med Assoc J.* 2006;174(11):1563–1569. doi:10.1503/cmaj.051679

29. Dirven H, Vist GE, Bandhakavi S, et al. Performance of preclinical models in predicting drug-induced liver injury in humans: a systematic review. *Sci Rep.* 2021;11(1):6403. doi:10.1038/s41598-021-85708-2

30. Pirmohamed M, James S, Meakin S, et al. Adverse drug reactions as cause of admission to hospital: prospective analysis of 18,820 patients. *Br Med J.* 2004;329(7456):15–19. doi:10.1136/bmj.329.7456.15

31. US Food and Drug Administration. *Preventable Adverse Drug Reactions: A Focus on Drug Interactions. ADRs: Prevalence and incidence.* 2019. https//www.fda.gov/Drugs/DevelopmentApprovalProcess/DevelopmentResources/DrugInteractionsLabeling/ucm110632.htm#ADRs%20Prevalence%20and%20Incidence

32. Lazarou J, Pomeranz BH, Corey PN. Incidence of adverse drug reactions in hospitalized patients. *JAMA.* 1998;279(15):1200. doi:10.1001/jama.279.15.1200

33. Bouvy JC, De Bruin ML, Koopmanschap MA. Epidemiology of adverse drug reactions in Europe: a review of recent observational studies. *Drug Saf.* 2015;38(5):437–453. doi:10.1007/s40264-015-0281-0

34. van Meer PJK, Kooijman M, Gispen-de Wied CC, Moors EHM, Schellekens H. The ability of animal studies to detect serious post marketing adverse events is limited. *Regul*

Toxicol Pharmacol. 2012;64(3):345–349. doi:10.1016/j.yrtph.2012.09.002

35. Bailey J, Thew M, Balls M. Predicting human drug toxicity and safety via animal tests: Can any one species predict drug toxicity in any other, and do monkeys help? *ATLA Altern to Lab Anim.* 2015;43(6):393–403. doi:10.1177/026119291504300607
36. Bailey J, Thew M, Balls M. An analysis of the use of animal models in predicting human toxicology and drug safety. *ATLA Altern to Lab Anim.* 2014;42(3):181–199. doi:10.1177/026119291404200306
37. Bailey J, Thew M, Balls M. An analysis of the use of dogs in predicting human toxicology and drug safety. *ATLA Altern to Lab Anim.* 2013;41(5):335–350.
38. Olson H, Betton G, Robinson D, et al. Concordance of the toxicity of pharmaceuticals in humans and in animals. *Regul Toxicol Pharmacol.* 2000;32(1):56–67. doi:10.1006/rtph.2000.1399
39. Matthews RAJ. Medical progress depends on animal models – doesn't it? *J R Soc Med.* 2008;101(2):95–98. doi:10.1258/jrsm.2007.070164
40. Monticello TM, Jones TW, Dambach DM, et al. Current nonclinical testing paradigm enables safe entry to First-in-Human clinical trials: the IQ consortium nonclinical to clinical translational database. *Toxicol Appl Pharmacol.* 2017;334:100–109. doi:10.1016/j.taap.2017.09.006
41. Clark M. Prediction of clinical risks by analysis of preclinical and clinical adverse events. *J Biomed Inform.* 2015;54:167–173. doi:10.1016/j.jbi.2015.02.008
42. Clark M, Steger-Hartmann T. A big data approach to the concordance of the toxicity of pharmaceuticals in animals and humans. *Regul Toxicol Pharmacol.* 2018;96:94–105. doi:10.1016/j.yrtph.2018.04.018
43. Bailey J, Balls M. Recent efforts to elucidate the scientific validity of animal-based drug tests by the pharmaceutical industry, protesting lobby groups, and animal welfare organisations. *BMC Med Ethics.* 2019;20(1):1–7. doi:10.1186/s12910-019-0352-3

44. Sievers S, Wieschowski S, Strech D. Investigator brochures for Phase I/II trials lack information on the robustness of preclinical safety studies. *Br J Clin Pharmacol*. 2020;1–9. doi:10.1111/bcp.14615
45. Wieschowski S, Chin WWL, Federico C, Sievers S, Kimmelman J, Strech D. Preclinical efficacy studies in investigator brochures: Do they enable risk–benefit assessment? *PLoS Biol*. 2018;16(4):e2004879. doi:10.1371/journal.pbio.2004879
46. Harris R. *Rigor Mortis*. Basic Books; 2017
47. Yarborough M, Bredenoord A, D'Abramo F, et al. The bench is closer to the bedside than we think: uncovering the ethical ties between preclinical researchers in translational neuroscience and patients in clinical trials. *PLoS Biol*. 2018;16(6):1–9. doi:10.1371/journal.pbio.2006343
48. Van Norman GA. Limitations of animal studies for predicting toxicity in clinical trials: Is it time to rethink our current approach? *JACC Basic to Transl Sci*. 2019;4(7):845–854. doi:10.1016/j.jacbts.2019.10.008
49. Editorial. Follow the yellow brick road. *Nat Rev Drug Discov*. 2003;2(3):167. doi:10.1038/nrd1057
50. Archibald K. Animal research is an ethical issue for humans as well as for animals. *J Anim Ethics*. 2018;8(1):1. doi:10.5406/janimalethics.8.1.0001
51. Harris R. Drugs that work in mice often fail when tried in people. *National Public Radio* clip. 2017. https://www.npr.org/sections/health-shots/2017/04/10/522775456/drugs-that-work-in-mice-often-fail-when-tried-in-people?t=1603725158790&t=1603789668763

Part Three: Breaking free

CHAPTER 8: THE POTENTIAL OF A HUMAN CELL
1. Skloot R. *The Immortal Life of Henrietta Lacks*. Picador; 2010
2. Associated Press in College Park Maryland. Henrietta Lacks' estate sues drug company that sold her cells. *Guardian*. October 4, 2021. https://www.theguardian.com/business/2021/oct/04/

henrietta-lacks-estate-sues-pharmaceutical-cells.
3. Nguyen N, Nguyen W, Nguyenton B, et al. Adult human primary cardiomyocyte-based model for the simultaneous prediction of drug-induced inotropic and pro-arrhythmia risk. *Front Physiol.* 2017;8. doi:10.3389/fphys.2017.01073
4. Watkins PB. Drug safety sciences and the bottleneck in drug development. *Clin Pharmacol Ther.* 2011;89(6):788–790. doi:10.1038/clpt.2011.63
5. Onakpoya IJ, Heneghan CJ, Aronson JK. Post-marketing withdrawal of 462 medicinal products because of adverse drug reactions: a systematic review of the world literature. *BMC Med.* 2016;14(1):1–11. doi:10.1186/s12916-016-0553-2
6. Baker M. A living system on a chip. *Nature.* 2011;471:661-665.
7. van Esbroeck ACM, Janssen APA, Cognetta AB, et al. Activity-based protein profiling reveals off-target proteins of the FAAH inhibitor BIA 10-2474. *Science.* 2017;356(6342):1084–1087. doi:10.1126/science.aaf7497
8. Fassbender M. New insight into BIA 10-2474 clinical trial shines light on preclinical testing. *Outsourcing Pharma.* June 18, 2017. https://www.outsourcing-pharma.com/Article/2017/06/19/BIA-10-2474-clinical-trial-update-shines-light-on-preclinical-testing?utm_source=copyright&utm_medium=OnSite&utm_campaign=copyright
9. University of Leiden. Experimental drug BIA 10-2474 de-activates proteins in human nerve cells. *Science Daily.* June 9, 2017. https://www.sciencedaily.com/releases/2017/06/170609092626.htm
10. Monmaney T. A triumph in the war against cancer. *Smithson Mag.* May 2011. https://www.smithsonianmag.com/science-nature/a-triumph-in-the-war-against-cancer-1784705/#LSCO5XW8eWmIsjJ1.99
11. Pray L. Gleevec: the breakthrough in cancer treatment. *Nat Educ.* 2008;1(1):37.
12. Bullen CK, Hogberg HT, Bahadirli-Talbott A, et al. Infectability of human BrainSphere neurons suggests neurotropism of

SARS-CoV-2. *ALTEX*. 2020;37(4):665–671. doi:10.14573/altex.2006111
13. Marshall M. Covid and the brain: researchers zero in on how damage occurs. *Nat*. July 7, 2021. https://www.nature.com/articles/d41586-021-01693-6
14. Kang I, Smirnova L, Kuhn J, Hogberg H, Kleinstreuer N, Hartung T. COVID-19 – prime time for microphysiological systems, as illustrated for the brain. *ALTEX*. 2021;38(4):535–549. doi:10.14573/altex.2110131
15. Takahashi K, Tanabe K, Ohnuki M, et al. Induction of pluripotent stem cells from adult human fibroblasts by defined factors. *Cell*. 2007;131(5):861–872. doi:10.1016/j.cell.2007.11.019
16. Skardal A, Aleman J, Forsythe S, et al. Drug compound screening in single and integrated multi-organoid body-on-a-chip systems. *Biofabrication*. 2020;12(2):25017. doi:10.1088/1758-5090/ab6d36
17. Low LA, Mummery C, Berridge BR, Austin CP, Tagle DA. Organs-on-chips: into the next decade. *Nat Rev Drug Discov*. 2021;20(5):345–361. doi:10.1038/s41573-020-0079-3
18. McCarthy C, Brewington JJ, Harkness B, Clancy JP, Trapnell BC. Personalised CFTR pharmacotherapeutic response testing and therapy of cystic fibrosis. *Eur Respir J*. 2018;51(6). doi:10.1183/13993003.02457-2017
19. Gallagher J. *'Organs-on-chips' wins design award*. BBC. June 23, 2015. https://www.bbc.co.uk/news/health-33237990.
20. Hamilton G. *Body Parts on a Chip*. TedX; 2013. https://www.ted.com/talks/geraldine_hamilton_body_parts_on_a_chip?language=en
21. Huh D, Matthews BD, Mammoto A, Montoya-Zavala M, Hsin HY, Ingber DE. Reconstituting organ-level lung functions on a chip. *Science*. 2010;328(5986):1662–1668. doi:10.1126/science.1188302
22. Huh D, Leslie DC, Matthews BD, et al. A human disease model of drug toxicity – induced pulmonary edema in a lung-on-a-chip microdevice. *Sci Transl Med*. 2012;4(159):159ra147.

doi:10.1126/scitranslmed.3004249
23. DARPA. *Microphysiological Systems*. 2010. https://www.darpa.mil/program/microphysiological-systems
24. Low LA, Tagle DA. Microphysiological systems ('organs-on-chips') for drug efficacy and toxicity testing. *Clin Transl Sci*. 2017;10(4):237–239. doi:10.1111/cts.12444
25. US Senate Committee on Appropriations. *Hearing on FY2017 National Institutes of Health Budget Request*. Testimony of Dr Francis Collins. 2016. https://www.appropriations.senate.gov/hearings/hearing-on-fy2017-national-institutes-of-health-budget-request
26. Ingber D. *Human Biology: An Exploration of Organs on Chips*. (Webinar) 2021. https://webinars.liebertpub.com/e/HumanBiologyAnExplorationofOrgansonChips
27. Ingber DE. Is it time for Reviewer 3 to request human organ chip experiments instead of animal validation studies? *Adv Sci*. 2020;7(22):1–15. doi:10.1002/advs.202002030
28. Kleinstreuer N, Holmes A. Harnessing the power of microphysiological systems for COVID-19 research. *Drug Discov Today*. 2021;26(11):2496–2501. doi:10.1016/j.drudis.2021.06.020
29. Si L, Bai H, Rodas M, et al. A human-airway-on-a-chip for the rapid identification of candidate antiviral therapeutics and prophylactics. *Nat Biomed Eng*. 2021;5(8):815–829. doi:10.1038/s41551-021-00718-9
30. Brownell L. Repurposing approved drugs for COVID-19 at an accelerated pace. 2020. https://wyss.harvard.edu/news/repurposing-approved-drugs-for-covid-19-at-an-accelerated-pace/
31. Jang K-J, Mehr AP, Hamilton GA, et al. Human kidney proximal tubule-on-a-chip for drug transport and nephrotoxicity assessment. *Integr Biol*. 2013;5(9):1119–1129. doi:10.1039/c3ib40049b
32. Cohen A, Ioannidis K, Ehrlich A, et al. Mechanism and reversal of drug-induced nephrotoxicity on a chip. *Sci Transl Med*. 2021;13(582). doi:10.1126/scitranslmed.abd6299

33. Jeffay N. Israelis create cancer drug without animal tests, by using human-simulating chip. *Times of Israel*. March 8, 2021. https://www.timesofisrael.com/israelis-create-cancer-drug-using-human-simulating-chip-instead-of-animal-tests/
34. Cookson C, Kuchler H, Miller J. How science is getting closer to a world without animal testing. *Financial Times*. August 14, 2022. https://www.ft.com/content/7c35e08a-4931-4401-b27e-acabf974bff8
35. Barrile R, van der Meer AD, Park H, et al. Organ-on-chip recapitulates thrombosis induced by an anti-CD154 monoclonal antibody: translational potential of advanced microengineered systems. *Clin Pharmacol Ther*. 2018;104(6):1240–1248. doi:10.1002/cpt.1054
36. Bavli D, Prill S, Ezra E, et al. Real-time monitoring of metabolic function in liver-on-chip microdevices tracks the dynamics of mitochondrial dysfunction. *Proc Natl Acad Sci*. 2016;113(16):E2231–E2240. doi:10.1073/pnas.1522556113
37. Ewart L, Apostolou A, Briggs SA, et al. Performance assessment and economic analysis of a human Liver-Chip for predictive toxicology. *Commun Med*. 2022;2(154):1–16. doi:10.1038/s43856-022-00209-1
38. Hartung T, de Vries R, Hoffmann S, et al. Toward good in vitro reporting standards. *ALTEX*. 2019;36(1):3–17. doi:10.14573/altex.1812191

CHAPTER 9: UNLEASHING THE POWER OF COMPUTERS

1. Archives IT. Interview with Professor Denis Noble. 2022. https://archivesit.org.uk/interviews/professor-denis-noble/
2. Carney S. Denis Noble discusses his career in computational biology. *Drug Discov Today: BIOSILICO*. 2004;2(4):135–137. doi:10.1016/S1741-8364(04)02414-X
3. Watkins PB. DILIsym: quantitative systems toxicology impacting drug development. *Curr Opin Toxicol*. 2020;23–24:67–73. doi:10.1016/j.cotox.2020.06.003
4. Smith B, Rowe J, Watkins PB, et al. Mechanistic investigations support liver safety of ubrogepant. *Toxicol Sci*. 2020;177(1):84–

93. doi:10.1093/toxsci/kfaa093

5. Passini E, Britton OJ, Lu HR, et al. Human in silico drug trials demonstrate higher accuracy than animal models in predicting clinical pro-arrhythmic cardiotoxicity. *Front Physiol.* 2017;8:1–15. doi:10.3389/fphys.2017.00668

6. Onakpoya IJ, Heneghan CJ, Aronson JK. Post-marketing withdrawal of 462 medicinal products because of adverse drug reactions: a systematic review of the world literature. *BMC Med.* 2016;14(1):1–11. doi:10.1186/s12916-016-0553-2

7. National Centre for the Replacement, Refinement and Reduction of Animals in Research. International 3Rs prize awarded for computer modelling that predicts human cardiac safety better than animal studies. March 12, 2018. https://nc3rs.org.uk/news/international-3rs-prize-awarded-computer-modelling-predicts-human-cardiac-safety-better-animal

8. Passini E, Zhou X, Trovato C, Britton OJ, Bueno-Orovio A, Rodriguez B. The virtual assay software for human in silico drug trials to augment drug cardiac testing. *J Comput Sci.* 2021;52:101202. doi:10.1016/j.jocs.2020.101202

9. Konduri PR, Marquering HA, van Bavel EE, Hoekstra A, Majoie CBLM. In-silico trials for treatment of acute ischemic stroke. *Front Neurol.* 2020;11:1–8. doi:10.3389/fneur.2020.558125

10. INSIST Consortium. *In Silico Clinical Trials for Acute Ischemic Stroke.* YouTube; 2020. https://www.youtube.com/watch?v=run-UWnPbHA&t=154s

11. Luechtefeld T, Marsh D, Rowlands C, Hartung T. Machine learning of toxicological big data enables read-across structure activity relationships (RASAR) outperforming animal test reproducibility. *Toxicol Sci.* 2018;165(1):198–212. doi:10.1093/toxsci/kfy152

12. Leichman AK. No animals harmed in new faster and cheaper chip drug tests. *Isr 21C.* December 16, 2021. https://www.israel21c.org/no-animals-harmed-in-new-chip-drug-tests-that-are-faster-and-cheaper/

13. Cookson C, Kuchler H, Miller J. How science is getting closer to a world without animal testing. *Financial Times.* August

14, 2022. https://www.ft.com/content/7c35e08a-4931-4401-b27e-acabf974bff8.
14. Van Norman GA. Limitations of animal studies for predicting toxicity in clinical trials: Part 2: potential alternatives to the use of animals in preclinical trials. *JACC Basic to Transl Sci*. 2020;5(4):387–397. doi:10.1016/j.jacbts.2020.03.010
15. Si L, Bai H, Rodas M, et al. A human-airway-on-a-chip for the rapid identification of candidate antiviral therapeutics and prophylactics. *Nat Biomed Eng*. 2021;5(8):815–829. doi:10.1038/s41551-021-00718-9
16. Parrish M. Why animal models are on the brink of extinction. *PharmaManuf.* May 2021. https://www.pharmamanufacturing.com/articles/2021/the-next-phase-of-drug-creation/
17. Stub ST. The secret Israeli sauce in Pfizer's new drug mix. *Times of Israel*. May 18, 2021. https://www.timesofisrael.com/spotlight/the-secret-israeli-sauce-in-pfizers-new-drug-mix/.
18. CompBioMed Collaboration. *About CompBioMed*. 2022. https://www.compbiomed.eu/about/
19. Weaving T. UCL-led team wins time on world's most powerful computer. December 5, 2022. https://www.compbiomed.eu/ucl-led-team-wins-time-on-worlds-most-powerful-computer/
20. Kohl P, Noble D. Systems biology and the virtual physiological human. *Mol Syst Biol*. 2009;5(292):1–6. doi:10.1038/msb.2009.51
21. Barrett A. Face to face with Denis Noble. *Oxford Sci*. February 1, 2022. https://oxsci.org/face-to-face-with-denis-noble/
22. Archibald K, Drake T, Coleman R. Barriers to the uptake of human-based test methods, and how to overcome them. *ATLA Altern to Lab Anim*. 2015;43(5):301–308. doi:10.1177/026119291504300504

CHAPTER 10: SOMETHING NEW, SOMETHING OLD
1. Pound P, Britten N, Morgan M, et al. Resisting medicines: a synthesis of qualitative studies of medicine taking. *Soc Sci Med*. 2005;61(1):133–155. doi:10.1016/j.socscimed.2004.11.063
2. Alliance for Human Relevant Science. *Accelerating the Growth*

of Human Relevant Life Sciences in the United Kingdom. A White Paper by the Alliance for Human Relevant Science. 2020. https//www.humanrelevantscience.org/wp-content/uploads/Accelerating-the-Growth-of-Human-Relevant-Sciences-in-the-UK_2020-final.pdf

3. Royal College of Physicians and British Pharmacological Society joint working party. *Personalised Prescribing: Using Pharmacogenomics to Improve Patient Outcomes.* 2022. https://www.rcp.ac.uk/projects/outputs/personalised-prescribing-using-pharmacogenomics-improve-patient-outcomes

4. National Human Genome Research Institute. *The Human Genome Project.* 2022. https://www.genome.gov/human-genome-project

5. Owen-Williams R. *Leading Causes of Death, UK: 2001 to 2018.* 2020. https://www.ons.gov.uk/peoplepopulationandcommunity/healthandsocialcare/causesofdeath/articles/leadingcausesofdeathuk/2001to2018

6. Pistollato F, Cavanaugh SE, Chandrasekera PC. A human-based integrated framework for Alzheimer's disease research. *J Alzheimer's Dis.* 2015;47(4):857–868. doi:10.3233/JAD-150281

7. Cavanaugh SE, Pippin JJ, Bernard N. Animal models of Alzheimer disease: historical pitfalls and a path forward. *ALTEX.* 2014;31(3):279–302. doi:10.14573/altex.1310071

8. Duyckaerts C, Potier MC, Delatour B. Alzheimer disease models and human neuropathology: similarities and differences. *Acta Neuropathol.* 2008;115(1):5–38. doi:10.1007/s00401-007-0312-8

9. Geerts H. Of mice and men. Bridging the translational disconnect in CNS drug discovery. *CNS Drugs.* 2009;23(1):915–926. https://doi.org/10.2165/11310890-000000000-00000

10. Hood L. Alzheimer's disease: my personal journey and search for a cure. *4sightHealth.* June 22, 2021. https://www.4sighthealth.com/alzheimers-disease-my-personal-journey-and-search-for-a-cure/

11. Bredeson D. *The End of Alzheimer's*. Penguin Random House; 2017
12. Rosenberg A, Ngandu T, Rusanen M, et al. Multidomain lifestyle intervention benefits a large elderly population at risk for cognitive decline and dementia regardless of baseline characteristics: The FINGER trial. *Alzheimer's Dement*. 2018;14(3):263–270. doi:10.1016/j.jalz.2017.09.006
13. Toups K, Hathaway A, Gordon D, et al. Precision medicine approach to Alzheimer's disease: successful pilot project. *J Alzheimer's Dis*. 2022;88(4):1411–1421. doi:10.3233/JAD-215707
14. Hood L. How technology, big data, and systems approaches are transforming medicine. *Res Technol Manag*. 2019;62(6):24–30. doi:10.1080/08956308.2019.1661077
15. Delude C. The details of disease (deep phenotyping). *Nature*. 2015;527:14–15
16. Price ND, Magis AT, Earls JC, et al. A wellness study of 108 individuals using personal, dense, dynamic data clouds. *Nat Biotechnol*. 2017;35(8):747–756. doi:10.1038/nbt.3870
17. National Institutes of Health. The future of health begins with you. NIH 'All of us' webpage. 2022. https://allofus.nih.gov/
18. Institute for Functional Medicine. *Functional Medicine: A Clinical Model to Address Chronic Disease and Promote Well-Being*. 2021. https://assets.website-files.com/60ff1f83fd29292e5a29f56d/62a38ebcbfbe162c6ed4c5faIFM_Functional_Medicine_Clinical_Model.pdf
19. Twin Health. *Whole Body Digital Twin*. 2021. https://www.usa.twinhealth.com/
20. Cedersund G. *Digital Twin Launch Event*. 2019. https://www.youtube.com/watch?v=MvWPHM7wWV4
21. Lindahl M. The goal is for everyone to have a digital twin. *Affärsstaden*. 2021. https://affarsstaden.se/esb-article/malet-ar-att-alla-ska-ha-en-digital-tvilling/
22. CompBioMed Collaboration. *Virtual Humans*. 2023. https://www.compbiomed.eu/general-public/
23. CompBioMed Collaboration. *CompBioMed Virtual*

Humans Film. 2019. https://www.youtube.com/channel/UCUiIfmesH_psiArXT3xcppA

24. Gregory A. NHS cancer patients to get pioneering genetic test to find best treatments. *Guardian.* January 31, 2022. https://www.theguardian.com/society/2022/jan/31/nhs-cancer-patients-to-get-pioneering-genetic-test-to-find-best-treatments

25. Mundasad S. Cancer: huge DNA analysis uncovers new clues. BBC. April 21, 2022. https://www.bbc.co.uk/news/health-61177584.

26. Smedley D, Smith KR, Martin A, et al. 100,000 genomes pilot on rare-disease diagnosis in health care – preliminary report. *N Engl J Med.* 2021;385(20):1868–1880. doi:10.1056/NEJMoa2035790

27. Oncology Think Tank. We must find ways to detect cancer much earlier. *Sci Am.* January 8, 2021. https://www.scientificamerican.com/article/we-must-find-ways-to-detect-cancer-much-earlier/

28. Raza A. *The First Cell and the Human Costs of Pursuing Cancer to the Last.* Basic Books; 2019

29. Raza A. *21st Century Innovations in Alternatives to Animals in Biomedical Research: Congressional panel on the Humane Research & Testing Act.* 2021. https://mailchi.mp/cc299895613b/shh1wdpt7w-538419#congress

30. NHS. Detecting cancer early. 2022. https://www.nhs-galleri.org

31. Kartal E, Schmidt TSB, Molina-E, et al. A faecal microbiota signature with high specificity for pancreatic cancer. *Gut Microbiota.* 2022:1–14. doi:10.1136/gutjnl-2021-324755

32. Rothwell PM. *Prevention of Stroke and Vascular Dementia.* 2013. https://www.youtube.com/watch?v=WQMH5rWrSTk

33. Wynick A. Meet the man who stops 10,000 strokes a year. *Oxford Mail.* March 12, 2014. https://www.oxfordmail.co.uk/news/11069058.meet-man-stops-10-000-strokes-year-videos/

34. Medical Research Council. *Annual Report and Accounts 2017/18.* 2018. https://mrc.ukri.org/publications/browse/annual-report-and-accounts-2017-18/

35. Unwin D, Delon C, Unwin J, Tobin S, Taylor R. What predicts drug-free type 2 diabetes remission? Insights from an 8-year general practice service evaluation of a lower carbohydrate diet with weight loss. *BMJ Nutr Prev Heal.* 2023:e000544. doi:10.1136/bmjnph-2022-000544
36. O'Hare R. First volunteers on COVID-19 human challenge study leave quarantine. 2021. https://www.imperial.ac.uk/news/218294/first-volunteers-covid-19-human-challenge-study/
37. Lopez Bernal J, Andrews N, Gower C, et al. Effectiveness of the Pfizer-BioNTech and Oxford-AstraZeneca vaccines on Covid-19 related symptoms, hospital admissions, and mortality in older adults in England: test negative case-control study. *BMJ.* 2021:n1088. doi:10.1136/bmj.n1088
38. Thwaites RS, Sanchez Sevilla Uruchurtu A, Siggins MK, et al. Inflammatory profiles across the spectrum of disease reveal a distinct role for GM-CSF in severe COVID-19. *Sci Immunol.* 2021;6(57):eabg9873. doi:10.1126/sciimmunol.abg9873
39. Mundasad S. Covid blood protein offers clues for treatments. BBC. March 11, 2021. https://www.bbc.co.uk/news/health-56352128
40. Cavanna F, Muller S, de la Fuente LA, et al. Microdosing with psilocybin mushrooms: a double-blind placebo-controlled study. *Transl Psychiatry.* 2022;12(1):307. doi:10.1038/s41398-022-02039-0
41. Rootman JM, Kiraga M, Kryskow P, et al. Psilocybin microdosers demonstrate greater observed improvements in mood and mental health at one month relative to non-microdosing controls. *Sci Rep.* 2022;12(1):11091. doi:10.1038/s41598-022-14512-3
42. Burt T, Young G, Lee W, et al. Phase 0/microdosing approaches: time for mainstream application in drug development? *Nat Rev Drug Discov.* 2020;19(11):801–818. doi:10.1038/s41573-020-0080-x
43. Gray M, Lagerberg T, Dombrádi V. Equity and value in 'Precision Medicine.' *New Bioeth.* 2017;23(1):87–94. doi:10.1080/20502877.2017.1314891

44. Brothers KB, Rothstein MA. Ethical, legal and social implications of incorporating personalized medicine into healthcare. *Futur Med.* 2015;12(1):43–51.
45. Forouzanfar MH, Afshin A, Alexander LT, et al. Global, regional, and national comparative risk assessment of 79 behavioural, environmental and occupational, and metabolic risks or clusters of risks, 1990–2015: a systematic analysis for the Global Burden of Disease Study 2015. *Lancet.* 2016;388(10053):1659–1724. doi:10.1016/S0140-6736(16)31679-8
46. Gaitskell K. Personalised medicine approaches to screening and prevention. *New Bioeth.* 2017;23(1):21–29. doi:10.1080/2050 2877.2017.1314884
47. Capewell S, Lloyd-Williams F. The role of the food industry in health: Lessons from tobacco? *Br Med Bull.* 2018;125(1):131–143. doi:10.1093/bmb/ldy002
48. Willet, W. *Summary report of the Eat-Lancet Commission. Healthy Diets From Sustainable Food Systems.* 2019. https://eatforum.org/content/uploads/2019/07/EAT-Lancet_Commission_Summary_Report.pdf
49. Coote JH, Joyner MJ. Is precision medicine the route to a healthy world? *Lancet.* 2015;385(9978):1617. doi:10.1016/S0140-6736(15)60786-3
50. Wilkinson R, Marmot M. (eds.) *Social Determinants of Health: The Solid Facts.* World Health Organization. 2003. https://apps.who.int/iris/handle/10665/326568

Part Four: The struggle to move forward

CHAPTER 11: REGULATORY DYSFUNCTION

1. European Commission. *Report from the Commission to the European Parliament and the Council on the Implementation of Directive 2010/63/EU on the Protection of Animals Used for Scientific Purposes in the Member States of the European Union.* 2020. https://eur-lex.europa.eu/legal-content/EN/TXT/PDF/?uri=CELEX:52020DC0015&from=EN

2. Busquet F, Kleensang A, Rovida C, Herrmann K, Leist M, Hartung T. New European Union statistics on laboratory animal use – what really counts! *ALTEX*. 2020;37(2):167–186. doi:10.14573/altex.2003241
3. Animals in Science Regulation Unit. *The Harm–Benefit Analysis Process: New Project Licence Applications. Advice Note 05/2015.* 2015. https://assets.publishing.service.gov.uk/government/uploads/system/uploads/attachment_data/file/660238/Harm_Benefit_Analysis__2_.pdf
4. Home Office. *Animals in Science Regulation Unit: Annual Reports 2019 to 2021.* 2022. https://assets.publishing.service.gov.uk/government/uploads/system/uploads/attachment_data/file/1112529/14.148_HO_Scientific_Procedures_ARA__web_.pdf
5. Schuppli C. Decisions about the use of animals in research: ethical reflection by ethics committee members. *Anthrozoos*. 2011;24(4):409–425.
6. Garner JP. The mouse in the room: the critical distinction between regulations and ethics. In: Beauchamp T, DeGrazia D, eds. *Principles of Animal Research Ethics*. Oxford University Press; 2020
7. Schuppli C, Fraser D. Factors influencing the effectiveness of research ethics committees. *J Med Ethics*. 2007;33(5):294–301.
8. Jones-Engel L. Opinion: hold animal use committees accountable for their failures. *The Scientist*. July 27, 2022. https://www.the-scientist.com/news-opinion/opinion-hold-animal-use-committees-accountable-for-their-failures-70278
9. Gluck JP. *Voracious Science and Vulnerable Animals*. University of Chicago Press; 2016
10. Brønstad A, Newcomer CE, Decelle T, Everitt JI, Guillen J, Laber K. Current concepts of harm–benefit analysis of animal experiments – report from the AALAS–FELASA Working Group on Harm–Benefit Analysis – Part 1. *Lab Anim*. 2016;50(1_suppl):1–20. doi:10.1177/0023677216642398
11. Davies G, Golledge H, Hawkins P, Rowland A, Smith J, Wolfensohn S. *Review of Harm–Benefit Analysis in the Use of*

Animals in Research. 2017. https://assets.publishing.service.gov.uk/government/uploads/system/uploads/attachment_data/file/675002/Review_of_harm_benefit_analysis_in_use_of_animals_18Jan18.pdf

12. Pound P, Nicol CJ. Retrospective harm benefit analysis of pre-clinical animal research for six treatment interventions. *PLoS One.* 2018;13(3):e0193758. doi:10.1371/journal.pone.0193758
13. Lyons D. *The Politics of Animal Experimentation.* Palgrave Macmillan; 2013
14. Russell W, Burch R. *The Principles of Humane Experimental Technique.* Methuen; 1959
15. National Centre for the Replacement, Refinement and Reduction of Animals in Research. *The 3Rs.* 2023. https://nc3rs.org.uk/who-we-are/3rs
16. Herrmann K, Flecknell P. Severity classification of surgical procedures and application of health monitoring strategies in animal research proposals: a retrospective review. *ATLA Altern to Lab Anim.* 2018;46(5):273–289. doi:10.1177/026119291804600508
17. Herrmann K, Flecknell P. Retrospective review of anesthetic and analgesic regimens used in animal research proposals. *ALTEX.* 2019;36(1):65–80. doi:10.14573/altex.1804011
18. Herrmann K, Flecknell P. The application of humane endpoints and humane killing methods in animal research proposals: a retrospective review. *ATLA Altern to Lab Anim.* 2018;46(6):317–333. doi:10.1177/026119291804600606
19. Lilley E, Armstrong R, Clark N, et al. Refinement of animal models of sepsis and septic shock. *Shock.* 2015;43(4):304–316. doi:10.1097/SHK.0000000000000318
20. Lilley E, Hawkins P, Jennings M. A 'road map' toward ending severe suffering of animals used in research and testing. *ATLA Altern to Lab Anim.* 2014;42(4):267–272. doi:10.1177/026119291404200408
21. Herrmann K. Refinement on the way towards replacement: Are we doing what we can? In: Herrmann K, Jayne K, eds.

Animal Experimentation: Working Towards a Paradigm Change. Brill; 2019
22. Home Office. *Annual Statistics of Scientific Procedures on Living Animals Great Britain 2021.* 2022. https://assets.publishing.service.gov.uk/government/uploads/system/uploads/attachment_data/file/1085383/annual-statistics-scientific-procedures-living-animals-2021_v8.pdf
23. Silverwood J. RSPCA chief quits group after losing confidence in regulation of animal testing. *Vet Times.* February 8, 2022. https://www.vettimes.co.uk/news/rspca-chief-quits-group-after-losing-confidence-in-regulation-of-animal-testing/
24. Marshall LJ, Constantino H, Seidle T. Phase-in to phase-out – targeted, inclusive strategies are needed to enable full replacement of animal use in the European Union. *Animals.* 2022;12:1–18
25. European Commission. *2019 Report on the Statistics on the Use of Animals for Scientific Purposes in the Member States of the European Union in 2015–2017.* 2020. https://www.eumonitor.eu/9353000/1/j9vvik7m1c3gyxp/vl5wgdto76xz
26. Taylor K, Alvarez LR. An estimate of the number of animals used for scientific purposes worldwide in 2015. *Altern Lab Anim.* 2019;47(5–6):196–213. doi:10.1177/0261192919899853
27. Grimm D. How many mice and rats are used in US labs? Controversial study says more than 100 million. *Science.* January 12, 2021. https://www.science.org/content/article/how-many-mice-and-rats-are-used-us-labs-controversial-study-says-more-100-million
28. United States Department of Agriculture. *Annual Report Animal Usage by Fiscal Year (2018).* 2020. https://www.aphis.usda.gov/aphis/ourfocus/animalwelfare/sa_obtain_research_facility_annual_report/ct_research_facility_annual_summary_reports
29. European Union. European Union Directive 2010/63/EU of the European Parliament and of the Council of 22 September 2010 on the protection of animals used for scientific purposes. *Off J Eur Union.* 2010;276.
30. Alliance for Human Relevant Science. *Bringing Back the*

Human: Transitioning from Animal Research to Human-Relevant Science in the UK. 2022. https://www.humanrelevantscience.org/all-party-parliamentary-group/bringing-back-the-human-transitioning-from-animal-research-to-human-relevant-science-in-the-uk/

31. UK Parliament. Ban commercial breeding for laboratories. Implement reform to approve & use NAMs. *Hansard.* 2023. https://hansard.parliament.uk/commons/2023-01-16/debates/CE7E0DD2-CD4E-4C47-A58D-983D6E2BC128/CommercialBreedingForLaboratories

32. Hutchinson I, Owen C, Bailey J. Modernizing medical research to benefit people and animals. *Animals.* 2022;12(9):1–12. doi:10.3390/ani12091173

33. Innovate UK. *A Non-Animal Technologies Roadmap for the UK: Advancing Predictive Biology.* 2015. https://assets.publishing.service.gov.uk/government/uploads/system/uploads/attachment_data/file/474558/Roadmap_NonAnimalTech_final_09Nov2015.pdf

34. Hobson-West P. What kind of animal is the 'Three Rs'? *Altern to Lab Anim.* 2009;37(2_suppl):95–99. doi:10.1177/026119290903702S11

35. Balls M. On the replacement of animal testing: yesterday, today, and tomorrow. 2019. https://www.youtube.com/watch?v=IdQ9tuHrYVM&t=3s

36. Balls M. It's time to reconsider the principles of humane experimental technique. *Altern Lab Anim.* 2020;48(1):40–46. doi:10.1177/0261192920911339

37. Swaters D, van Veen A, van Meurs W, Turner JE, Ritskes-Hoitinga M. A history of regulatory animal testing: What can we learn? *Altern to Lab Anim.* 2022:026119292211180. doi:10.1177/02611929221118001

38. Steele, D. Vanda Pharmaceuticals president receives award for work to prevent FDA dog tests. *ScienMag.* December 12, 2019. https://scienmag.com/vanda-pharmaceuticals-president-receives-award-for-work-to-prevent-fda-dog-tests/

39. Vanda Pharmaceuticals. Vanda Pharmaceuticals provides

update on tradipitant development program. *Biospace*. August 31, 2020. https://www.biospace.com/article/releases/vanda-pharmaceuticals-provides-update-on-tradipitant-development-program/

40. Patterson EA, Whelan MP, Worth AP. The role of validation in establishing the scientific credibility of predictive toxicology approaches intended for regulatory application. *Comput Toxicol*. 2021;17:100144. doi:10.1016/j.comtox.2020.100144

41. Mondou M, Maguire S, Pain G, et al. Envisioning an international validation process for new approach methodologies in chemical hazard and risk assessment. *Environ Adv*. 2021;4:100061. doi:10.1016/j.envadv.2021.100061

42. van der Zalm AJ, Barroso J, Browne P, et al. A framework for establishing scientific confidence in new approach methodologies. *Arch Toxicol*. 2022;96:2865–2879. doi:10.1007/s00204-022-03365-4

43. Turner J, Pound P, Owen C, et al. Incorporating new approach methodologies into regulatory preclinical pharmaceutical safety assessment. *Altex, Altern to Anim Exp*. 2023. doi:https://doi.org/10.14573/altex.2212081

44. National Centre for the Replacement, Refinement and Reduction of Animals in Research. Opportunities for use of one species for longer-term toxicology testing during drug development: a cross-industry evaluation. 2021. https://nc3rs.org.uk/sites/default/files/documents/Opportunities for use of a single species in drug development – Q%26A.pdf

45. Camp C, Bennett A, Beale A. Human models for human drug development. *Drug Discov World*. October 14, 2020. https://www.ddw-online.com/human-models-for-human-drug-development-7463-202010/

46. Archibald K, Drake T, Coleman R. Barriers to the uptake of human-based test methods, and how to overcome them. *ATLA Altern to Lab Anim*. 2015;43(5):301–308. doi:10.1177/026119291504300504

47. Congress.Gov. *H.R.2565 – FDA Modernization Act of 2021*.

2021. https://www.congress.gov/bill/117th-congress/house-bill/2565/
48. FDA. Advancing alternative methods at FDA. 2023. https://www.fda.gov/science-research/about-science-research-fda/advancing-alternative-methods-fda

CHAPTER 12: LOCKED IN

1. Alliance for Human Relevant Science. *Bringing Back the Human: Transitioning from Animal Research to Human-Relevant Science in the UK*. 2022. https://www.humanrelevantscience.org/all-party-parliamentary-group/bringing-back-the-human-transitioning-from-animal-research-to-human-relevant-science-in-the-uk/
2. Biotechnology and Biological Sciences Research Council, UK Research and Innovation. *BBSRC Strategic Delivery Plan 2022–2025*. 2022. https://www.ukri.org/wp-content/uploads/2022/09/BBSRC-010922-StrategicDeliveryPlan2022.pdf
3. Charity Commission for England and Wales. Animal Free Research UK: income/expenditure. 2023. https://register-of-charities.charitycommission.gov.uk/charity-search/-/charity-details/5027694/financial-history
4. Charity Commission for England and Wales. The Humane Research Trust: income/expenditure. 2023. https://register-of-charities.charitycommission.gov.uk/charity-search/-/charity-details/267779/financial-history
5. Marshall LJ, Triunfol M, Seidle T. Patient-derived xenograft vs. organoids: a preliminary analysis of cancer research output, funding and human health impact in 2014–2019. *Animals*. 2020;10(1923). doi:10.3390/ani10101923
6. Nuwer R. US agency seeks to phase out animal testing. *Nat Index*. November 4, 2022. https://www.nature.com/articles/d41586-022-03569-9
7. Frank J. Technological lock-in, positive institutional feedback, and research on laboratory animals. *Struct Chang Econ Dyn*. 2005;16(4):557–575. doi:10.1016/j.strueco.2004.11.001
8. Krebs C, Camp C, Constantino H, et al. Proceedings of

a workshop to address animal methods bias in scientific publishing. *ALTEX.* 2022. doi:10.14573/altex.2210211

9. Del Pace L, Viviani L, Straccia M. Researchers and their experimental models: a pilot survey in the context of the European Union health and life science research. *Animals.* 2022;12(20):2778. doi:10.3390/ani12202778

10. Kuhn T. *The Structure of Scientific Revolutions.* University of Chicago Press; 1962

11. Lynn L. *Pathological Consensus.* 2022. https://www.youtube.com/watch?v=_JWb1ruvEdM

12. Preuss T. What animal models do to us. National Institutes of Health Neuroscience Series Seminar. 2016. https://videocast.nih.gov/summary.asp?Live=17982&bhcp=1

13. Gluck JP. *Voracious Science and Vulnerable Animals.* University of Chicago Press; 2016

14. Pound P, Ram R. Are researchers moving away from animal models as a result of poor clinical translation in the field of stroke? An analysis of opinion papers. *BMJ Open Sci.* 2020;4(1):e100041. doi:10.1136/bmjos-2019-100041

15. Greenhalgh T, Robert G, Macfarlane F, Bate P, Kyriakidou O. Diffusion of innovations in service organizations: systematic review and recommendations. *Milbank Q.* 2004;82(4):581–629. doi:10.1111/j.0887-378X.2004.00325.x

16. Ritskes-Hoitinga M, Pound P. The role of systematic reviews in identifying the limitations of preclinical animal research, 2000–2022: Part 2. *J R Soc Med.* 2022;115(6):231–235. doi:10.1177/01410768221100970

17. Birke L, Arluke A, Michael M. *The Sacrifice. How Scientific Experiments Transform Animals and People.* Purdue University Press; 2007

18. Latour B. Drawing things together. In: Lynch M, Woolgar S, eds. *Representations in Scientific Practice.* MIT Press; 1990

19. Pound P. Animal models: problems and prospects. In: Fischer B, ed. *The Routledge Handbook of Animal Ethics.* Taylor & Francis; 2019:584

20. Pound P, Ritskes-Hoitinga M. Is it possible to overcome issues

of external validity in preclinical animal research? Why most animal models are bound to fail. *J Transl Med.* 2018;16(1):304. doi:10.1186/s12967-018-1678-1

21. Bottini AA, Hartung T. Food for thought ... on the economics of animal testing. *ALTEX.* 2009;26(1):3–16. doi:10.14573/altex.2009.1.3

22. The Business Research Company. Animal testing and non-animal alternative testing market players invest in novel technologies as per The Business Research Company's *Animal Testing and Non-Animal Alternative Testing Global Market Report 2022.* 2022. https://www.globenewswire.com/en/news-release/2022/01/26/2373712/0/en/Animal-Testing-And-Non-Animal-Alternative-Testing-Market-Players-Invest-In-Novel-Technologies

23. Market Statsville Group. Animal model market is projected to reach USD 2,676.8 million by 2030. August 22, 2022. https://www.openpr.com/news/2711108/animal-model-market-is-projected-to-reach-usd-2-676-8-million

24. LaFollette H, Shanks N. *Brute Science: Dilemmas of Animal Experimentation.* Routledge; 1997

25. Balls M. It's time to reconsider the principles of humane experimental technique. *Altern Lab Anim.* 2020;48(1):40–46. doi:10.1177/0261192920911339

26. McGlacken R. (Not) knowing and (not) caring about animal research: an analysis of writing from the Mass Observation Project. *Sci Technol Stud.* 2021;35(3). doi.org/10.23987/sts.102496

27. Arluke A. Trapped in a guilt cage. *New Sci.* 1992;134(1815):33–35

28. Lynch M. Sacrifice and the transformation of the animal body into a scientific object. Laboratory culture and ritual practice in the neurosciences. *Soc Stud Sci.* 1988;18:267

29. Fortes M. Preface. In: Bourdillon M, Fortes M, eds. *Sacrifice.* Academic Press; 1980

30. Lienhardt G. *Divinity and Experience: The Religion of the Dinka.* Clarendon Press; 1961

31. Lang S. *Challenges.* Springer; 1997

32. Ioannidis JPA. Why science is not necessarily self-correcting. *Perspect Psychol Sci.* 2012;7(6):645–654. doi:10.1177/1745691612464056
33. Vazire S, Holcombe AO. Where are the self-correcting mechanisms in science? *Rev Gen Psychol.* 2022;26(2):212–223. doi:10.1177/10892680211033912

CHAPTER 13: DEATH THROES AND BIRTH PANGS

1. McKie R. Scientists split as genetics lab scales down animal tests. *Guardian.* June 9, 2019. https://www.theguardian.com/science/2019/jun/09/genetics-laboratory-scales-down-animal-tests
2. Sample I. UK's leading mouse genetics centre faces closure. *Guardian.* June 20, 2019. https://www.theguardian.com/science/2019/jun/20/uk-mouse-genetics-centre-faces-closure-threatening-research
3. Innovate UK. *A Non-Animal Technologies Roadmap for the UK: Advancing Predictive Biology.* 2015. https://assets.publishing.service.gov.uk/government/uploads/system/uploads/attachment_data/file/474558/Roadmap_NonAnimalTech_final_09Nov2015.pdf
4. Home Office, Department for Business, Innovation & Skills, and Department of Health. *Working to Reduce the Use of Animals in Scientific Research: Delivery Report.* 2015. https://assets.publishing.service.gov.uk/government/uploads/system/uploads/attachment_data/file/417441/Delivery_Report_2015.pdf
5. BioIndustry Association and Medicines Discovery Catapult. *State of the Discovery Nation 2018 and the Role of the Medicines Discovery Catapult.* 2018. https//md.catapult.org.uk/FlipBuilder/mobile/index.html
6. National Research Council. *Toxicity Testing in the 21st Century: A Vision and a Strategy.* The National Academies Press; 2007. https://nap.nationalacademies.org/catalog/11970/toxicity-testing-in-the-21st-century-a-vision-and-a
7. Food and Drug Administration. *Advancing Alternative*

Methods at FDA. 2023. https://www.fda.gov/science-research/about-science-research-fda/advancing-alternative-methods-fda

8. Interagency Coordinating Committee on the Validation of Alternative Methods. *A Strategic Roadmap for Establishing New Approaches to Evaluate the Safety of Chemicals and Medical Products in the United States.* 2018. https://ntp.niehs.nih.gov/iccvam/docs/roadmap/iccvam_strategicroadmap_january2018_document_508.pdf

9. Alliance for Human Relevant Science. *Accelerating the Growth of Human Relevant Life Sciences in the United Kingdom. A White Paper by the Alliance for Human Relevant Science.* 2020. https//www.humanrelevantscience.org/wp-content/uploads/Accelerating-the-Growth-of-Human-Relevant-Sciences-in-the-UK_2020-final.pdf

10. Wheeler A. *Directive to Prioritize Efforts to Reduce Animal Testing.* 2019. https://www.epa.gov/sites/default/files/2019-09/image2019-09-09-231249.txt

11. DiMasi JA, Grabowski HG, Hansen RW. Innovation in the pharmaceutical industry: new estimates of R&D costs. *J Health Econ.* 2016;47:20–33. doi:10.1016/j.jhealeco.2016.01.012

12. Franzen N, van Harten WH, Retèl VP, Loskill P, van den Eijnden-van Raaij J, IJzerman M. Impact of organ-on-a-chip technology on pharmaceutical R&D costs. *Drug Discov Today.* 2019;24(9):1720–1724. doi:10.1016/j.drudis.2019.06.003

13. Ewart L, Apostolou A, Briggs SA, et al. Performance assessment and economic analysis of a human Liver-Chip for predictive toxicology. *Commun Med.* 2022;2(154):1–16. doi:10.1038/s43856-022-00209-1

14. Hutchinson I, Owen C, Bailey J. Modernizing medical research to benefit people and animals. *Animals.* 2022;12(9):1–12. doi:10.3390/ani12091173

15. Centre for Economics and Business Research. *The Economic Impact of the UK's New Approach Methodologies Sector – A CEBR Report for Animal Free Research UK.* 2021. https://cebr.com/reports/the-economic-impact-of-the-uks-new-approach-

methodologies-sector-a-cebr-report-for-animal-free-research-uk/
16. The Business Research Company. Animal testing and non-animal alternative testing market players invest in novel technologies as per The Business Research Company's *Animal Testing and Non-Animal Alternative Testing Global Market Report 2022.* 2022. https://www.globenewswire.com/en/news-release/2022/01/26/2373712/0/en/Animal-Testing-And-Non-Animal-Alternative-Testing-Market-Players-Invest-In-Novel-Technologies
17. Allied Market Research. *Animal Model Market: Global Opportunity Analysis and Industry Forecast 2021–2030.* 2021. https://www.alliedmarketresearch.com/animal-model-market-A07946
18. Allied Market Research. *Non-Animal Alternative Testing Market: Global Opportunity Analysis and Industry Forecast, 2021–2030.* 2022. https://www.alliedmarketresearch.com/non-animal-alternative-testing-market-A25675
19. Ipsos Mori. *Public Attitudes to Animal Research in 2018.* 2019. https://www.ipsos.com/sites/default/files/ct/news/documents/2019-05/18-040753-01_ols_public_attitudes_to_animal_research_report_v3_191118_public.pdf
20. Animal Free Research. Poll: clear majority of Britons want end to animal testing in UK labs. March 23, 2021. https://www.animalfreeresearchuk.org/poll-clear-majority-of-britons-want-end-to-animal-testing-in-uk-labs/
21. Savanta ComRes. *RSPCA – Animal Testing Poll.* 2022. https://savanta.com/knowledge-centre/poll/rspca-animal-testing-poll/
22. Gallup. *Moral Issues.* 2023. https://news.gallup.com/poll/1681/moral-issues.aspx
23. Savanta ComRes. *Cruelty Free Europe – Animal Testing in the EU.* 2020. https://savanta.com/knowledge-centre/poll/cruelty-free-europe-animal-testing-in-the-eu/
24. Groff K, Bachli E, Lansdowne M, Capaldo T. Review of evidence of environmental impacts of animal research and testing. *Environ – MDPI.* 2014;1(1):14–30. doi:10.3390/environments1010014
25. International Union for the Conservation of Nature.

Long-tailed Macacque. 2021. https://www.iucnredlist.org/species/12551/221666136

26. Ipsos Mori. *Openness in Animal Research: The Public's Views on Openness and Transparency in Animal Research.* 2013. https://concordatopenness.org.uk/wp-content/uploads/2017/04/openness-in-animal-r.pdf

27. Understanding Animal Research. *Concordat on Openness in Animal Research in the UK.* 2014. http://concordatopenness.org.uk/wp-content/uploads/2017/04/Concordat-Final-Digital.pdf.

28. Workman P. Being open about animal research. 2014. https://www.icr.ac.uk/blogs/the-drug-discoverer/page-details/being-open-about-animal-research

29. FRAME (Fund for the Replacement of Animals in Medical Experiments). *What Is Section 24 of the Animals Scientific Procedure Act 1986?* 2023. https://frame.org.uk/the-issue/what-is-section-24-of-the-animals-scientific-procedure-act-1986/

30. UK Parliament. Ban commercial breeding for laboratories. Implement reform to approve & use NAMs. *Hansard.* 2023. https://hansard.parliament.uk/commons/2023-01-16/debates/CE7E0DD2-CD4E-4C47-A58D-983D6E2BC128/CommercialBreedingForLaboratories

31. Animal Research Tomorrow. *Basel Declaration.* 2022. https://animalresearchtomorrow.org/en/basel-declaration

32. Animal Research Tomorrow. *Animal Research 2020 and Beyond.* Online conference. 13 November, 2020. https://animalresearchtomorrow.org/en/conferences/animal-research-2020-and-beyond

33. Eggel M, Neuhaus CP, Grimm H. Reevaluating benefits in the moral justification of animal research: a comment on necessary conditions for morally responsible animal research. *Cambridge Q Healthc Ethics.* 2020;29(1):131–143. doi:10.1017/S0963180119000860

34. Grasser LR, Stammen R. What is ethical animal research? A scientist and veterinarian explain. *Conversat.* November 23, 2022. https://theconversation.com/what-is-ethical-animal-

research-a-scientist-and-veterinarian-explain-190876
35. Pound P, Ritskes-Hoitinga M. Is it possible to overcome issues of external validity in preclinical animal research? Why most animal models are bound to fail. *J Transl Med.* 2018;16(1):304. doi:10.1186/s12967-018-1678-1
36. Steckler T, Macleod M. Drug discovery and preclinical drug development – Have animal studies really failed? *BMJ Open Sci Blog.* 2019. https://blogs.bmj.com/openscience/2019/02/22/drug-discovery-and-preclinical-drug-development-have-animal-studies-really-failed/
37. Editorial. EPA drops target date to end mammalian toxicity testing by 2035. *Altex, Altern to Anim Exp.* December 20, 2021. https://www.altex.org/index.php/altex/announcement/view/358
38. Deutsche Forschungsgemeinschaft. *Position Paper on Securing Efficient Biomedical Research While Maintaining the Highest Animal Welfare Standards.* 2022. https://www.dfg.de/download/pdf/dfg_im_profil/gremien/senat/tierexperimentelle_forschung/tp_biomedizinische_forschung_de.pdf
39. Medical Research Council, UK Research and Innovation. *MRC Strategic Delivery Plan 2022–2025.* 2022. https://www.ukri.org/wp-content/uploads/2022/09/MRC-200922-MRCStratrgicDeliveryPlan.pdf
40. Pound P, Ram R. Are researchers moving away from animal models as a result of poor clinical translation in the field of stroke? An analysis of opinion papers. *BMJ Open Sci.* 2020;4(1):e100041. doi:10.1136/bmjos-2019-100041
41. Lacreuse A, Bennett A, Detmer A. Expanding Alzheimer's research with primates could overcome the problem with treatments that show promise in mice but don't help humans. *Conversat.* August 31, 2022. https://theconversation.com/expanding-alzheimers-research-with-primates-could-overcome-the-problem-with-treatments-that-show-promise-in-mice-but-dont-help-humans-188207
42. Vitek MP, Araujo JA, Fossel M, et al. Translational animal models for Alzheimer's disease: an Alzheimer's Association

Business Consortium Think Tank. *Alzheimer's Dement Transl Res Clin Interv.* 2020;6(1). doi:10.1002/trc2.12114

43. Animal Research Tomorrow. *Animal Research Tomorrow (Former Basel Declaration Society): Annual Report 2020.* 2021. https://animalresearchtomorrow.org/en/annual-reports

44. Archibald K, Tsaioun K, Kenna JG, Pound P. Better science for safer medicines: the human imperative. *J R Soc Med.* 2018;111(12):433–438. doi:10.1177/0141076818812783

45. Balls M. It's time to reconsider the principles of humane experimental technique. *Altern Lab Anim.* 2020;48(1):40–46. doi:10.1177/0261192920911339

46. Camp C, Bennett A, Beale A. Human models for human drug development. *Drug Discov World.* October 14, 2020. https://www.ddw-online.com/human-models-for-human-drug-development-7463-202010/

47. van der Zalm AJ, Barroso J, Browne P, et al. A framework for establishing scientific confidence in new approach methodologies. *Arch Toxicol.* 2022;96:2865–2879. doi:10.1007/s00204-022-03365-4

48. Sohn E. What if we didn't have to test new drugs on animals? *NeoLife.* October 2022

49. Drug Discovery World. *Using Microfluidics in Drug Discovery.* 2022. https://www.ddw-online.com/using-microfluidics-in-drug-discovery-19436-202210/

50. Ingber DE. Is it time for Reviewer 3 to request human organ chip experiments instead of animal validation studies? *Adv Sci.* 2020;7(22):1–15. doi:10.1002/advs.202002030

51. Balls M. It's time to include harm to humans in harm–benefit analysis – but how to do it, that is the question. *ATLA Altern to Lab Anim.* 2021;49(5):182–196. doi:10.1177/02611929211062223

52. Kickbusch L, Buckett K. *Implementing Health in All Policies: Adelaide 2010.* 2010. https://www.sahealth.sa.gov.au/wps/wcm/connect/0ab5f18043aee450b600feed1a914d95/implementinghiapadel-sahealth-100622.pdf

53. Transitie Proefdiervrije Innovatie (Transition to Animal Free

Innovations). Review of TPI 2018–2020. 2020. https://www.transitieproefdiervrijeinnovatie.nl/documenten/rapporten/20/11/11/review-of-tpi

54. Forest D. 'I wish I could make all animal experimenters feel what I felt when I had my aha moment.' *Anim People Forum*. August 2020. https://animalpeopleforum.org/2020/08/10/i-wish-i-could-make-all-animal-experimenters-feel-what-i-felt-when-i-had-my-aha-moment/

55. von Aulock S. Engagement of scientists with the public and policymakers to promote alternative methods. *ALTEX*. 2022:543–559. doi:10.14573/altex.2209261

56. Alliance for Human Relevant Science. *Bringing Back the Human: Transitioning from Animal Research to Human-Relevant Science in the UK*. 2022. https://www.humanrelevantscience.org/all-party-parliamentary-group/bringing-back-the-human-transitioning-from-animal-research-to-human-relevant-science-in-the-uk/

57. European Union. European Union Directive 2010/63/EU of the European Parliament and of the Council of 22 September 2010 on the protection of animals used for scientific purposes. *Off J Eur Union*. 2010;276.

58. Pound P, Blaug R. Transparency and public involvement in animal research. *ATLA Altern to Lab Anim*. 2016;44:167–173

59. Brunt MW, Weary DM. Public consultation in the evaluation of animal research protocols. *PLoS One*. 2021;16:1–11. doi:10.1371/journal.pone.0260114

60. Mondou M, Maguire S, Pain G, et al. Envisioning an international validation process for new approach methodologies in chemical hazard and risk assessment. *Environ Adv*. 2021;4:100061. doi:10.1016/j.envadv.2021.100061

61. DeGrazia D, Sebo J. Necessary conditions for morally responsible animal research. *Cambridge Q Healthc Ethics*. 2015;24(4):420–430. doi:10.1017/S0963180115000080

62. Linzey A, Linzey C, Peggs K. *Normalising the Unthinkable: The Ethics of Using Animals in Experiments*. Oxford Centre for Animal Ethics; 2015

63. Archibald K, Coleman R, Drake T. Replacing animal tests to improve safety for humans. In: Herrmann K, Jayne K, eds. *Animal Experimentation: Working Towards a Paradigm Change.* Brill; 2019:417–442. doi:10.1163/9789004391192_019
64. Archibald K. Animal research is an ethical issue for humans as well as for animals. *J Anim Ethics.* 2018;8(1):1. doi:10.5406/janimalethics.8.1.0001

INDEX

A

academia: 28–31, 187–189
Académie des Sciences: 6
airway-on-a-chip: 124, 137
All Party Parliamentary Group: 211
allocation concealment: 40–41
ALS Therapy Development Institute: 42
Alzheimer's disease: xii, 33, 65, 85, 88, 144–148, 207–208
American Cancer Society: 23
amodiaquine: 137
amyotrophic lateral sclerosis: 42
anaesthesia: 5, 12, 17
Animal Free Research UK: 91, 124, 171, 176, 184, 211
Animal Research Tomorrow: 89, 204
Animal Welfare (Sentience) Act 2022: xvi
Animal Welfare and Ethical Review Body (AWERB): 165–167, 170–171
Animals in Science Regulation Unit (ASRU): 164–165, 169
Animals (Scientific Procedures) Act 1986: 31, 164–165, 203
antibiotics: 90, 101
applied research: 32, 70
Archibald, Kathy: 100–101, 109, 198, 208–209
Aristotle: 4
arrhythmia: 132–133
artificial intelligence: 134–141
aspirin: 84, 109
Association for the Advancement of Medicine by Research: 15
asthma: 89, 122
Aston University: 71, 107
AT132: 103
atherosclerosis: 69
atorvastatin: 103
autism: 136
autopsies: 4, 76
Akhtar, Aysha: 65, 72

INDEX

B

Bacon, Francis: 11
Bailey, Jarrod: 91–92, 106–107, 176
Baillie, Kenneth: 156
Balls, Michael: 91, 172, 209
Bar-Zohar, Danny: 136
Basel Declaration: 89, 204
basic research: 32, 70, 93,
Bates, Alan: 16
Bayliss, William: 17
beclomethasone: 73
Bentwich, Isaac: 135–136
Bernard, Claude: 3, 6–10, 14, 18–19, 58, 63, 76,
Bernard, Marie-Francoise: 3
BIA 10-2474: 98–99, 117
big data: 134–135
Biotechnology and Biological Sciences Research Council: 184
Birke, Lynda: 192
Blakemore, Colin: xii, 38
blinded outcome assessment: 40–41
Bracken, Michael: 37
Bradford, Daniel: 97
brain tumour: 181–183, 192–193
British Medical Association: 13, 15
British Medical Journal: xi, xvi, 22, 38
Brown Dog Affair: 16–17
Burch, Rex: 169
Burdon-Sanderson, John: 12–13
burns: 45–46, 210

C

CAMARADES: 38
cancer: xii, 23, 33, 43, 47–48, 53, 61–62, 86–88, 90–92, 109, 115–116, 118–119, 121, 125–126, 150–154, 185, 203
Cairns-Smith, Graham: 62
cardiovascular disease: 69, 86, 93, 102–103
Carney, Stephen: 130
Carvalho, Constança: 80
Case Western Reserve University: 68
Castle, William Ernest: 23
cat: 44, 109, 181
Cedersund, Gunnar: 149
Center for Alternatives to Animal Testing: 120, 204
Center for Contemporary Sciences: 65, 72
Center for Drug Evaluation and Research: 174, 178
Centre for Animals and Social Justice: 168
Centre for Economics and Business Research: 199
Charles River Laboratories: 33
chemotherapy: 87, 125, 153, 192–193
Cheng, Frances: 68, 82
cholesterol: 69, 103, 149
cisplatin: 125–126
Coalition for Medical Progress: xii, 31
Coleman, Bob (Robert): 73
Collège de France: 6–7
Collins, Francis: 108, 124

Columbia University: 47, 87
Commission on Human Medicines: 27
CompBioMed project: 139, 150
Concordat on Openness in Animal Research: xiii, 203–204
confounding: 42–44
Contopoulos-Ioannidis, Despina: 92
Covid-19: 33, 48, 54, 120, 124–125, 137, 156
Crohn's disease: 89
Cruelty to Animals Act 1876: 12–13, 17, 20
cystic fibrosis: xii, 121

D

Darwin, Charles: 9–10, 58–59
data cloud: 147
deep phenotyping: 147
Defense Advanced Research Projects Agency (DARPA): 123
dementia: xii, 85, 145, 154, 199
Descartes, René: 5
Deutsche Forschungsgemeinschaft: 207
diabetes: 33, 50, 72, 89, 105, 125, 127, 155–156
diaspirin: 102
digital twin: 149–150
Dines, Sarah: 170
Directive 2010/63/EU: 170, 179, 212–213
dog: 16–18, 20–21, 60, 87, 99, 102, 104, 109, 118–119, 174, 181

Donaldson, Henry: 24
Druker, Brian: 118–119
drug-induced liver injury (DILI): 117, 131
Dutch Burns Foundation: 210

E

Early Cancer Institute: 151
early detection: 145–147, 151–154, 159
Ebrahim, Shah: 37–38
Eggel, Matthias: 205
Ehler, Christian: 206
Einstein, Albert: 85
empagliflozin: 125
endotoxin: 45–46
enlimomab: 102
Environmental Protection Agency (EPA): xvii, 198, 206
European Commission: 197
European Medicines Agency: 197
European Parliament: xvii, 197, 206
evolutionary biology: xviii, 57–62, 64, 74
evolutionary theory: 10, 57–61, 66–67, 74, 107, 195
Exascale computer: 139

F

Faculté de Medicine: 14
false positives: 109, 153, 158
Farrar, Jeremy: 196
FDA Modernization Act 2022: 178, 198
Federal Food, Drug, and Cosmetic Act 1938: 178

Ferranti Mercury: 129
ferrets: 23
fialuridine: 101, 117
Fink, Mitchell: 46
first-in-human trials: 99, 157
Food and Drug Administration (FDA): 50, 100, 118, 123, 124, 132, 174–175, 178, 198
Fragile X Syndrome: 136
Frank, Joshua: 29, 185
functional medicine: 148
Fund for the Replacement of Animals in Medical Experiments (FRAME): 211
Futuyma, Douglas: 58

G

Galen: 4
Galleri test: 153
gangrene: 98
Garner, Joseph: 50
genetically modified animals: 31–32, 214
glaucoma: 68
gleevec: 109, 119
glivec: 119
Gluck, John: 30–31, 81–82, 167
gold standard: 176, 179, 185, 208
Grant, Jonathan: 79–80
GSK2193874: 123
guinea pig: 22–23, 73, 76–77, 96, 107, 109

H

Hackam, Daniel: 92
Hamilton, Geraldine: 122
Hannover Medical School: 49, 79
Harel, David: 138
Harlow, Harry: 81
harm–benefit analysis: 165, 167–168, 174, 205
Harries, Lorna: 60, 113–115, 119, 128
Hartung, Thomas: 120, 134–135
Harvard University: 23, 123, 137
Harvey, William: 5
Harwell Institute: 197, 207
Hebrew University: 125–126
HeLa cells: 116
Helpathon: 210
Henderson, Mark: xii
hepatitis B: 101
hepatocyte: 117
Herbert Irving Comprehensive Cancer Center: 47, 87
Herrmann, Kathrin: 164, 204, 206
Hickman, James: 126
high blood pressure: 46, 50, 158
Hippocrates: 4
HIV/AIDS: 89
Home Office: 165, 168, 170–171, 197, 212
homology: 10, 59, 195
Hood, Leroy: 145–148
Horn, Janneke: 77
Human Genome Project: 143
human-on-a-chip: 123, 128
human tissue: 66, 113, 116, 126, 182
Humane Society International: 170
Hunter, John: 18
Hutter, Otto: 129
hygiene: 89

I

Imperial College London: 52
in silico: xvi, 131–135, 139–141, 149, 158–159, 175, 185–186, 198
in vitro: 116–117, 127, 131, 138, 158, 175, 176, 177, 183, 185,186, 210
in vivo: 131
induced pluripotent stem cell (iPSC): 120
inflammation: 45–46, 47
inflammatory bowel disease: 49
Ingber, Donald: 122–124, 137
Innovate UK: 171
Institute for Cancer Research: 203
Institute for Systems Biology: 145
Institute of Psychiatry: 182
insulin: 73
International Medical Congress: 13, 15
investigator brochure: 108
Ioannidis, John: 43, 92
ivabradine: 131

J

Jackson Laboratory: 23, 33
JAX mice: 23
Johns Hopkins University: 120, 134

K

Keynes, John Maynard: 209
Khan, Raste: 97
Kingsford, Anna: 14
Knight, Andrew: 80
Koch, Robert: 8, 13

Krijnen, Paul: 210
Kuhn, Thomas: xviii, 186, 189–190

L

Laboratory Animals Bureau: 22
Lacks, Henrietta: 115–116
Lafollette, Hugh: 32, 58, 62, 74, 90, 94, 191
Lathrop, Abbie: 23
Latour, Bruno: 29
Leenaars, Cathalijn: 49, 79
Le Fort, Leon: 14
Leiden University: 99
leukaemia: 87–88, 90, 96, 109–110, 118, 152
Lind af Hageby, Lizzy: 17
lithium: 76–77
Little, Clarence Cook: 23
liver transplant: 102
London School of Medicine for Women: 17
lymphoma: 87
Lyons, Dan: 168

M

machine learning: 134–138
Maeterlinck, Maurice: 202
Magendie, François: 6–7, 19
magnetic resonance imaging (MRI): 157
major depressive disorder: 80, 89
Marshall, Lindsay: 92
Massachusetts Institute of Technology: 123
maternal deprivation: 81
Matthews, Robert: 71, 107

McKeown, Thomas: 89–90
Medical Research Committee: 21
Medical Research Council: 22, 26, 38, 155, 197, 207
Medicines and Healthcare products Regulatory Agency (MHRA): 98, 102, 175, 177–178
mental health: 48, 157
methotrexate: 49
Meyer, Adolf: 23–24
mice: xv, 21–24, 28, 31, 33, 42–43, 45–48, 60–63, 65–66, 68, 88, 99, 104, 115, 118, 138, 170, 172, 181, 196, 210
microbiome: 61, 128, 145, 147, 154, 158
microdosing: 157
Mintz, Beatrice: 31
MK3207: 132
monkeys: xi, xv, 30, 96, 99, 118
mortality rates: 34, 86, 90
motor neuron disease: 42, 89, 108
mouse: 21, 23, 31, 45–46, 48, 60–62, 100, 138, 197, 207
multiple organ failure: 98
multiple sclerosis: 49, 89, 96
myelodysplastic syndromes: 152

N

Nahmias, Yaakov: 126
National Centre for Replacement, Refinement and Reduction of Animals in Research (NC3Rs): 172, 184
National Institute for Biological Standards and Control: 98
National Institute for Medical Research: 22
National Institutes of Health: 25, 63, 92, 101, 108, 123, 124, 185
National Insurance Act 1911: 21–22
neuroprotective: 85
Nicol, Christine: 173
nimodipine: 77
Noble, Denis: 129–130, 132, 140
non-human primates: 54, 59, 81, 168, 181, 204
Northwick Park Hospital: 97
Nuremberg guidelines: 27

O

Oakley, David: 97
obesity: 44, 54, 86
Oldfield, Rob: 96
oligodendroglioma: 192
Oncology Think Tank: 152
organoid: xvi, 119–121, 127, 135–136, 196, 204
organ-on-a-chip: xvi, 121–128, 135, 198–199, 209
osteoarthritis: 49, 89, 120

P

Parkinson's disease: 49, 89, 98
Passini, Elisa: 132–133
Pasteur Institute: 9, 45
Pasteur, Louis: 13
pathological consensus: 186
patient-on-a-chip: 135–136
payback analysis: 93
penicillin: 73

Perlman, Robert: 60–62, 66
personalised medicine: 121, 139, 158
People for the Ethical Treatment of Animals (PETA): 82
Peters, Keith: 71
Petsko, Gregory: 110
pharmacogenomics: 143
phylogenetic scale: 64
Physiological Society: 13, 20
physiology: 6–8, 11, 68, 75, 124, 130, 194
Pilkington, Geoff: 181–184, 185, 187, 192
polio: xii, 73
Polymeropoulos, Mihael: 174–175
Porton Down: 26
Portsmouth Brain Tumour Research Centre: 183
post-mortem studies: 120, 157,
Preuss, Todd: 25, 63–65
prevention: 94, 141, 144, 150, 155, 195
primary cells: 114, 116–117
project licence: 165–166, 169, 171, 203, 212
psychiatry: 81, 182
public health: 65, 90, 105, 155, 158, 195, 198, 201
publication bias: 52
pulmonary oedema: 123

R

rabbits: 22–23, 76
Radboud University: 41, 188
random allocation: 40–41
Ram, Rebecca: 56
Ramé, Maria Louise: 7
Rand Corporation: 79
ranitidine: 73
rats: xv, 22–24, 28–29, 31, 44, 58, 63, 65, 68–69, 73, 99, 109, 118, 170, 181, 196
Raza, Azra: 47–48, 86–88, 140, 152–154, 193
Redelmeier, Donald: 92
reduction: 165, 169–170, 172, 205
refinement: 165, 169, 172
regulations: xviii, 163–180, 202, 212, 213
regulatory capture: 168, 179
Reines, Brandon: 75–77
replacement: 106, 133, 165, 170–172, 211, 212
repurposing: 78, 85, 124, 136–137
Research Defence Society: 18, 31, 182
rezulin: 127
rheumatoid arthritis: 47, 49, 89, 96
Rhodes Farm: 22
Riolan, Jean: 11
risk of bias: 39–42
Ritskes-Hoitinga, Merel: 41, 55–56, 188–189, 205
Roberts, Ian: 37
rofecoxib: 104,
Romanes, George John: 20
Rosenblueth, Arturo: 44
Rothwell, Peter: 93, 154–155
Royal Commission on Vivisection: 12

Royal Society: xii, 70–71
Royal Society of Medicine: 86
Rupke, Nicolaas: 15
Russell, Bill: 169

S

sacrifice: 192
Safer Medicines Trust: 101, 198
salbutamol: 73
sample size calculation: 41
Sandercock, Peter: 37
Sanger Institute: 196
sanitation: 89
SARS-CoV-2: 120, 137, 156
Schafer, Edward: 20
Schartau, Leisa: 17
sheep: x, xi
schizophrenia: 93
Schouten, Carola: 206
Scott, Sean: 42–43
Select Committee Inquiry on the use of Animals in Scientific Procedures 2002: 71
selfotel: 102
sepsis: 45, 89
serum: 114
Shanks, Niall: 32, 58, 62, 74, 90, 94, 191
sickle cell anaemia: 66
sildenafil: 78
Sina, Ibn: 11, 67
Sinclair, Upton: 191
Singer, Peter: 30
Skloot, Rebecca: 115
Speaking of Research: 207
species differences: 10–11, 18, 34, 55, 56–67, 100, 109, 132, 189, 195–196, 208
STI571: 118–119
standard animal: 21–22, 24
Stanford University: 43, 50, 92
stem cell: 120
Stratton, Mike: 196
stroke: ix, xii, 43, 46, 48–49, 50, 52, 55, 56, 77, 78, 83–85, 88, 94, 98, 102–103, 120, 133–134, 139, 154–155, 213
St Thomas Hospital Medical School: ix
supercomputer: 134, 139
SYRCLE: 38–39, 41
systems biology: xvi, 139–140, 145

T

tamoxifen: 109
telcagepant: 132
TGN1412: 96–99
thalidomide: 26–27
three Rs (3Rs): 165, 169, 172–173, 204–205
thrombosis: 126
tirilazad: 48, 102
tissue plasminogen activator (tPA): 78, 84–85
tissue repository: 152
torcetrapib: 103
tradipitant: 174–175
transgenic: 33, 65–66
transient ischaemic attack: 155
traumatic brain injury: 89
troglitazone: 105
Tsilidis, Konstantinos: 52

U

ubrogepant: 132
UK Medicines Act 1968: 27
ulcerative colitis: 139
Understanding Animal Research: xii, 31, 57, 72, 74, 202, 205
University College London: 129
University Medical Center Utrecht: 53
University of Amsterdam: 77
University of Bergen: 43
University of Bristol: xiii, 37–38, 142, 173
University of California: 46, 108
University of Cambridge: 71, 93
University of Chicago: 60
University of Edinburgh: 156
University of Exeter Medical School: 60, 113
University of Lisbon: 80
University of Nottingham: 91, 172
University of Oxford: 93, 130, 132, 154
University of Pittsburgh: 75
University of Portsmouth: 183
University of Winchester: 80
Utrecht University: 55, 189

V

vaccines: xii, 73, 90
van der Naald, Mira: 53
van Esbroeck, Annelot: 99
van Schie, Carine: 210
Vesalius, Andreas: 5
Vioxx: xiii, 104
Virchow, Rudolf: 13
virtual human: 137–140, 150

W

Walport, Mark: 74
Warren, H Shaw: 45–46
wearable sensors: 135, 149, 152
Weill Cornell Medicine: 110
Wellcome Trust: xii, 196
Wiener, Norbert: 44
Wilks, Samuel: 15
Wilson, Ryan: 97–98
Wistar Institute: 24
Wistar rat: 24, 28–29, 33, 63
Wooding, Steven: 93
World War: 21, 26, 90
Wyss Institute: 123–124, 137

Y

Yarborough, Mark: 108
Yerkes National Primate Research Center: 25, 63

Z

Zeller, Rolf: 204

This book is printed on paper from sustainable sources managed under the Forest Stewardship Council (FSC) scheme.

It has been printed in the UK to reduce transportation miles and their impact upon the environment.

For every new title that Troubador publishes, we plant a tree to offset CO_2, partnering with the More Trees scheme.

MORE TREES
LET'S PLANT A BILLION TREES

For more about how Troubador offsets its environmental impact, see www.troubador.co.uk/about/

ABOUT THE AUTHOR

Pandora Pound is Research Director at Safer Medicines Trust, a charity that aims to improve the safety of medicines by facilitating a transition to human biology-based drug development and testing. She has a PhD in the Sociology of Medicine and over two decades' experience of conducting research. In 2004, she transformed the debate on animal experiments as lead author of a landmark paper published in the British Medical Journal which led to a series of systematic reviews that ultimately exposed the scientific limitations of using animals in medical research. She has written numerous academic papers on the scientific drawbacks of using animals as models for humans. She is a Fellow of the Oxford Centre for Animal Ethics.